BEFORE YOU START READING, DOWNLOAD YOUR FREE BONUSES!

Click the link or scan the QR-code & access all the resources for FREE!

The Self-Sufficient Living Cheat Sheet
10 Simple Steps to Become More Self-Sufficient in 1 Hour or Less

How to restore balance and beauty to the environment around you… even if you live in a tiny apartment in the city.
Discover:

- **How to increase your income** by selling "useless" household items
- The environmentally friendly way to replace your car — invest in THIS special vehicle to **eliminate your carbon footprint**
- The secret ingredient to **turning your backyard into a thriving garden**
- 17+ different types of food scraps and 'waste' that you can use to feed your garden
- How to drastically **cut down on food waste** without eating less
- 4 natural products you can use to make your own eco-friendly cleaning supplies
- The simple alternative to 'consumerism' — the age-old method for **getting what you need without paying money for it**
- The 9 fundamental items you need to create a self-sufficient first-aid kit
- One of the top skills that most people are afraid of learning — and how you can master it effortlessly
- 3 essential tips for **gaining financial independence**

The Prepper Emergency Preparedness & Survival Checklist:

10 Easy Things You Can Do Right Now to Ready Your Family & Home for Any Life-Threatening Catastrophe

Natural disasters demolish everything in their path, but your peace of mind and sense of safety don't have to be among them. Here's what you need to know...

- Why having an emergency plan in place is so crucial and how it will help to keep your family safe
- How to stockpile emergency supplies intelligently and why you shouldn't overdo it
- How to store and conserve water so that you know you'll have enough to last you through the crisis
- A powerful 3-step guide to ensuring financial preparedness, no matter what happens
- A step-by-step guide to maximizing your storage space, so you and your family can have exactly what you need ready and available at all times
- Why knowing the hazards of your home ahead of time could save a life and how to steer clear of these in case of an emergency
- Everything you need to know for creating a successful evacuation plan, should the worst happen and you need to flee safely

101 Recipes, Tips, Crafts, DIY Projects and More for a Beautiful Low Waste Life

Reduce Your Carbon Footprint and Make Earth-Friendly Living Fun With This Comprehensive Guide

Practical, easy ways to improve your personal health and habits while contributing to a brighter future for yourself and the planet

Discover:

- **Simple customizable recipes for creating your own food, home garden, and skincare products**
- The tools you need for each project to successfully achieve sustainable living
- Step-by-step instructions for life-enhancing skills from preserving food to raising your own animals and forging for wild berries
- **Realistic life changes that reduce your carbon-footprint while saving you money**
- Sustainable crafts that don't require any previous knowledge or expertise
- Self-care that extends beyond the individual and positively impacts the environment
- **Essential tips on how to take back control of your life -- become self-sustained and independent**

First Aid Fundamentals
A Step-By-Step Illustrated Guide to the Top 10 Essential First Aid Procedures Everyone Should Know

Discover:

- **What you should do to keep this type of animal attack from turning into a fatal allergic reaction**
- Why sprains are more than just minor injuries, and how you can keep them from getting worse
- **How to make the best use of your environment in critical situations**
- The difference between second- and third-degree burns, and what you should do when either one happens
- Why treating a burn with ice can actually cause more damage to your skin
- When to use heat to treat an injury, and when you should use something cold
- **How to determine the severity of frostbite**, and what you should do in specific cases
- Why knowing this popular disco song could help you save a life
- The key first aid skill that everyone should know — **make sure you learn THIS technique the right way**

Food Preservation Starter Kit

10 Beginner-Friendly Ways to Preserve Food at Home | Including Instructional Illustrations and Simple Directions

Grocery store prices are skyrocketing! It's time for a self-sustaining lifestyle. Discover:

- **10 incredibly effective and easy ways to preserve your food for a self-sustaining lifestyle**
- The art of canning and the many different ways you can preserve food efficiently without any prior experience
- A glorious trip down memory lane to learn the historical methods of preservation passed down from one generation to the next
- **How to make your own pickled goods**: enjoy the tanginess straight from your kitchen
- Detailed illustrations and directions so you won't feel lost in the preservation process
- The health benefits of dehydrating your food and how fermentation can be **the key to a self-sufficient life**
- **The secrets to living a processed-free life** and saving Mother Earth all at the same time

Download All the resources by scanning the QR-Code below:

Backyard Beekeeping for Beginners 2022-2023

Step-By-Step Guide To Raise Your First Colonies in 30 Days With the Most Up-To-Date Information

Small Footprint Press

Table of Contents

Introduction

"The hum of the bees is the voice of the garden"—Elizabeth Lawrence.

Does being in nature bring you unparalleled joy and contentment? When you think about the world of bugs, insects, and other critters, do you find that you are constantly captivated, and you find yourself interested in learning more? Maybe you are concerned with what humanity is doing to the planet and want to explore the various ways in which you can give back to mother nature?

If you answered yes to any of these questions, becoming a beekeeper should be right up your alley!

Humans are hardwired to build, create, and work for things that give us joy. As we have become bogged down with office work, jobs that do not challenge us or stimulate our minds, that joy has begun to diminish. We spend far too much of our time sitting around doing menial tasks that do not have much of an effect on our world or our spirits. Nowadays, we find ourselves feeling disconnected from nature and not properly understanding how we can return to a rightful balance with the natural world around us.

It is understandable if you feel this way in today's world, but there are many things you can do to reconnect. One of the most powerful and impactful ways to give back to Mother Earth and build a fun new hobby for yourself at the same time is to start your very own beehive. Bees are nature's most precious asset, and without them, the world would truly never be the same again.

Becoming a beekeeper is an excellent way to fulfill the primal need that we talked about above. It gets you out and about and lets you create and do things for yourself rather than being reliant on others. It lets you become more self-sufficient and return to the land, rather than buying products from corporations that mass-produce the food in chain stores—no step of that process is great for the Earth.

Becoming a beekeeper is very rewarding, but it is not for the faint of heart or those following a whim. It takes time, sweat-inducing effort, and money, but the rewards are numerous. Not only are you getting a captivating and fun new way to pass the time, but you are also getting quality one-on-one interaction with nature and giving back to it at the same time. And of course, you get the reward of honey at the end. Store-bought honey is OK, but it cannot beat the quality or the satisfaction that comes from that first bite of honey you produced yourself.

Along with that sense of pride, beekeeping is a great pursuit for those who are interested in bees or other insects. Becoming a beekeeper will increase your knowledge of the small but vital part of the ecosystem. You will also get another bonus of being able to observe the little critters to your heart's content and see how they interact with the world around them.

Or maybe you want to get into beekeeping for reasons other than just giving yourself something to do other than work. Bees are a vital part of our lives, as they play a role in every single part of our planet's ecosystem. Around 90% of all plants in the world need to be cross-pollinated to survive, and in North America, bees are the most important movers in pollination (Home and Garden, 2021).

As is well known by now, bee species all over the world are dying at a truly alarming rate. It might be tempting to think of that as a "somewhere else" problem, but it's affecting us close to home as well. In Ontario, Canada, one common species of bumblebee was put on the endangered species list in 2009 and has not been seen in the wild since then (Wildlife Preservation Canada, 2021). And it is not the only bee species to be disappearing—between October 2018 and April 2019, the

University of Maryland reported that 40% of bee colonies in the United States died, making it the highest winter loss in 13 years (Woodward, 2019).

The reasons that the bees are dying are numerous and happen all over the globe:

- Climate change has created later growing seasons, so bees cannot get the pollen they need to survive early enough in the year.
- They have lost their natural habitat and spaces to city development, resulting in fewer spaces for bees to live.
- With large areas of land dedicated to growing a single kind of crop (monoculture), bees lack the plants and flowers that sustain them.

So, what do the bees dying mean to us? Remember, bees are the major source of pollination for both wild and cultivated plants. So, if all the bees were to die, plant life on our planet would stop growing. Or at least stop growing in the manner with which we are most familiar. We might be able to come up with artificial methods of pollination for crops, but replicating what nature has already perfected is never easy or cheap. What if that method was not cost-effective and plants that corporations could not make a profit from were no longer being pollinated? Would that leave us with a barren post-apocalyptic-style world?

Maybe things would not be that dire, but it is a concern all the same. Thinking about those kinds of things might be why you are interested in becoming a beekeeper.

The good news is that by providing a place for bees to live and be safe, you are also providing up to one acre of pollination per hive you run (Government of New Brunswick, 1996). That is a pretty good area of coverage and would contribute a lot to your local flora. Even small steps, done all together, impact the wider world and make it a better place.

Everything presented in this book will help guide you in your first steps toward your end goal—being a full-fledged apiarist. To make all the information as accessible as possible, everything that you read here will be presented straightforwardly and simply.

That way, you can get the best knowledge you can without getting bogged down by terms or information that a beginner does not need to know. The tips, tricks, and advice presented here are all gathered from beekeepers with years of experience and are designed to set any newbie up for success rather than failure.

As a company, Small Footprint Press is focused and determined on providing the best help possible. We want to make it easy for people to return to the land and live more sustainably for the long term.

We do this by doing in-depth research for all our books, with topics ranging from survival training to various self-sufficiency methods and prepping for disasters (natural or human-caused). We work our hardest to teach you how to care for the planet Earth and its inhabitants in the most holistic and mutually beneficial manners possible. Our team is made of dedicated and enthusiastic nature lovers, who altogether have over three decades of combined knowledge of outdoorsmanship and conservation. We can't wait to pass it all along to you.

Our goal is to help empower you to achieve the sustainable lifestyle that you want. Like you, we feel that giving back to our planet is crucial for our continued survival as a species. We full-heartedly believe that being outdoors (in a sustainable manner) makes us humans happy, and we want to share that happiness with you as well.

We know that the idea of starting a beehive can be intimidating, but we are here to help you take it one step at a time. This book is designed to help you know how to follow your path—both before starting on the road and during your first year. We emphasize building a proper knowledge base and understanding before getting started so that you can avoid as many pitfalls as possible.

By the time you are done with this book, you are going to be feeling much more confident in your knowledge and abilities. As long as you have the right knowledge and tools, getting started and succeeding in beekeeping is not as hard as you might imagine.

In this book, you will learn about:

- The benefits of joining a beekeeping club and the many ways that doing so will help make you a better apiarist.
- An accurate estimate of the time and money that you are going to be investing in your first year.
- What to look for in choosing the location of your beehive and why this is vital for your success.
- Preparing your hive for success and happiness, both for yourself and for your bees.
- How to install bees into your hive and set it up so everything goes swimmingly right from the start.
- Common mistakes and pitfalls that befall beekeepers in their first years (and how to avoid them).
- BONUS: Some basic information on how to turn beekeeping into some extra cash!

Enough with the dreaming of one day having your prized colony in your backyard.

It's a great time to get involved in beekeeping, and it will bring you tremendous joy as you give back to Mother Earth and help restore nature to its beauty and balance.

No matter what your current experience level is, with the right network and information, you can establish your very own successful backyard beehive in no time at all and with zero added stress.

Chapter 1:

Prepping for Success

It might seem like the first step you need to take in becoming a beekeeper is to get some bees, but that would be jumping the gun—by a lot. Before you rush off to the store or website to buy a thousand-dollar hive box and hundreds of little bees, it is better to have a base layer of knowledge to fall back on first. If you want to succeed, you first have to prep for success. If you do not, you are working toward disaster.

Good thing that we are going to be covering all bases of preparation. We will first look at something that bees are great at doing—making connections and working together. Before you get started on your own, it is best to make some new friends who are experienced beekeepers.

You can do this by joining a local or online beekeeping club, many of which offer perks beyond socializing. The people in these clubs are just as excited and passionate about bees as you are, and they want to see you succeed in your efforts. There are many different benefits of joining a local club, including training courses, sharing information and tips, gaining a mentor, and eventually allowing you to be a mentee. After all, one of the best ways to test your knowledge is to share it with someone else.

By being a part of a club, you will have a rich resource of knowledge and experience at your fingertips.

But it is good to go into a new group with at least a little knowledge as well. Reading this book, specifically this first chapter, is a great resource for getting some of that initial understanding. The second part of this chapter will go over some of the initial information that you should have before getting too heavily invested in the idea of beekeeping.

Beekeeping Clubs

Have you ever had a subject you were interested in and passionate about? One that you wanted to talk about to everyone all the time, but you were worried that you were boring or pushy? Does your interest in bees fit into those descriptions as well?

Well, if so, do not worry. There is a way to still talk nonstop about bees but not get annoyed looks from those around you. The solution is simple—join a beekeeping club. No more needing to cut yourself off or change the subject because all the people surrounding you will want to talk about bees too!

Like with all clubs or groups, there are numerous reasons to join your local beekeeping club, not the least of which is making new friends. Finding a group to join could be as easy as Googling "beekeeping clubs near me." There are also numerous groups, ranging from local clubs made up of the same ten people to national-level organizations, such as the American Beekeeping Federation and the Canadian Association of Professional Apiarists.

Why join a club? There are numerous benefits of doing so, the main one being that many clubs will offer training courses. These kinds of courses are usually heavily based on theory and so are geared toward beginner beekeepers, but they can also be good refreshers for more experienced apiarists or if it has been a while since you first learned the information. A good way to judge the training offered by your local clubs is to see how they teach—if they immediately go into the more complex aspects of beekeeping, it is probably not the course for you.

You want courses taught by knowledgeable people who teach straightforwardly, starting from the foundation and then building up.

Many people who are passionate about the subject may forget that they know things that are not common knowledge and so can use terms that mean nothing to you, which does not make it easy to learn anything. If you can, test out several different local clubs to see which has the best fit for you. If there is only one club in your area, and they teach like this, talk to the teacher and let them know that they are going too fast for any beginner. Hopefully, they will take your feedback in stride and adjust their lessons accordingly.

So, what kind of information should be included in an introduction to beekeeping? The first thing that should be taught is an understanding of the seasons and what needs to be done for your bees in each one. This part of the course should not have too much detail, but it should clearly state what happens in each season, such as the inactive periods that make up most of the year, the high-production time of early spring, and how to prepare your hives for the winter. This information should then be used to help put the rest of the course into the context of a timeframe.

Beekeeping courses should also be teaching you the practical and not just the theoretical parts of hive management. Many newcomers to the practice think that it is as easy as buying a hive and some bees, and then you are rewarded with honey, but that is not the case. For many, when they realize that there is more work involved in keeping a hive than they thought, they give up on their hives which defeats the purpose entirely.

The overview of the practical elements includes being upfront about the time and effort commitment you are making when purchasing a hive. Bees need to be fed nectar during the times of the year when there is not much natural food for them to collect. When preparing for winter, they are going to need even more nectar than normal to help them during the cold months.

Bees also need a lot of room to grow their colony, and their hives need regular inspections for pests, diseases, or if they are preparing for a swarm. Swarming is a very normal action and is the method of the bee colony reproducing or splitting itself into two colonies once it has grown too large for just one queen. Most often, swarming occurs in the spring, but it can happen later in the year as well, and so this is something that beekeepers need to watch out for. If your bees look like they are going to swarm, you will need to act sooner rather than later.

Since you need to be mindful of issues like these and more, it is recommended that inspections happen every week. Any course on beekeeping needs to be upfront about this because otherwise, they are creating a false expectation. As well, it should give you the tools and information that you need to look out for in these inspections. They should also discuss the risks of hive inspections, including getting stung, the heavy-lifting for when your hive is full of honey, the various weather conditions you will be working in, and the unavoidable reality that some bees will be squashed in the process.

These are the basics that should be covered in any beekeeping course you take. Of course, these lessons do not have to happen in person. Online courses are available as well, though the information presented will be more generalized. Going through a class on the basics of bees is a great idea—that way, you start getting an understanding of what beekeeping entails and seeing if it is for you.

It is recommended to take a more in-depth course in person at a local club. That way, the information will be based on the lived experience of the teacher. Another advantage of taking such a course at a local club is that these people know your area and what to expect from it in terms of weather and bee behavior.

Of course, the benefits of joining a club go beyond the training that it offers. A club is made up of more people than just one teacher, so by interacting with the other members, you are almost guaranteed to gain more information as they share their experiences and stories. While listening to everyone can be helpful and entertaining, the best way to take advantage of the expertise at your fingertips is to find a mentor.

Gaining a mentor gives you not only way more knowledge but also a person willing to help you as you set up your hive and guide you through the tricky parts. Having someone willing to come and see your hives and how they are doing gives you greater flexibility in how to care for your bees properly.

To find a mentor, you will first need to get to know everyone at your local club and determine who fits with you best. That fit needs to be not only in personality and temperament but also in project scope and goals. If you are just looking at working with one hive, you would not want someone who has experience with hundreds of hives at a time. By matching scale and end goals, you can also more easily create a collaboration project between the two of you, which sweetens the deal for your mentor. Mentors may also agree because teaching is a great way for them to solidify their knowledge, which you may experience if you stick around long enough.

Of course, beekeeping clubs are not just good for courses and mentorship. Like we said earlier, these organizations are great places to meet like-minded people who are interested in the same things you are. These people may become your new best friends as you work on your hives, and they will encourage your growth and push you to explore different methods of beekeeping you may not have heard of otherwise.

Most clubs do a lot to foster a sense of community, including having frequent meetings and get-togethers for everyone to catch up and share the latest news from their hives. Often these meetings can be themed around something of concern to the members, such as learning about the latest disease affecting hives in the country. Many clubs will also offer special interest events, such as honey tastings and workshops on brewing mead or making beeswax cloths.

So, now that you know the importance of joining a beekeeping club let us take a quick reminder about how to find one. Google is your friend in this search, especially if it takes you to the website of a national organization such as the American Beekeeping Federation of Bee Culture (names to search for both are at the end of the chapter). These web pages are great sources of information, both in figuring out the location of the closest beekeeping club to you and in signing up for additional resources such as monthly newsletters and online forums.

Many of these online groups also have ties to numerous other pages and organizations for you to check out. Additionally, they probably have links to their own and other social media accounts, which allows you to connect to bee lovers all over the world. Joining some of these online groups will let you expand your beekeeper friend group and gain insight from very different kinds of apiarists. This expands on several qualities seen in the local clubs, just on a much larger scale.

What Else Should I Know?

Now that we have looked at the benefits of joining a beekeeping club, we should cover some of the other things you need to consider before buying your first colony.

Something of the utmost importance that you need to know is the regulations around hives and bees in your area. Your beekeeping club can help you here, as well as help keep each other up to date with any new information or potential changes to the regulations.

But it is also important for you to do your due diligence and research to empower yourself and stay up to date on your own. The last thing you want is to have only part of the picture of regulations around honeybees, get your colony thriving and producing delicious honey, only to be shut down due to some rule you did not know.

Regulations and Laws

So, what kinds of regulations do you need to know? Well, the first and biggest one you need to know is whether or not your city even allows beekeeping at all. While some states and provinces have laws stating that beekeeping is legal, laws and regulations will still vary from city to city, so it is important to find your local laws.

The benefits of beekeeping are becoming more well known, and backyard beekeepers are a growing demographic. Because of this, most cities are accepting of this hobby, though they may have limits on the number and placements of your hives. Additionally, you may need to register each of your hives with the city. You can find this information by contacting your municipal bylaw office, looking on their website, or by asking people you know who keep bees. Generally, the requirements that are in place are easily followed and are not time-consuming to follow.

When finding out your cities requirements, you will need to check out the zoning laws, which will state whether or not bees are allowed either as of right (a legal term that means there is no permit required), allowed with a special permit, or not allowed outright. If it is written vaguely or confusingly, talk to your local zoning authority, who should be able to clear up any confusion you might have.

Aside from city regulation, you may also need to find out if there are any rules in place in your homeowner's association (HOA) in your neighborhood, which may differ from city ordinances. It is important when searching these rules to determine whether or not your HOA prohibits livestock (which can sometimes include bees) and what general rules pertain to keeping insects. If you plan to set up your hives as a side hustle, you will also need to check what your HOA states around home-based businesses.

You may be thinking that you need a nice big backyard to house a beehive. Many people could immediately assume that since they are living in an apartment and therefore have no yard, beekeeping is automatically not an option for them. But, there are ways to get around that hiccup. In the last few years, there has been a growing trend toward rooftop hives. If you want to set up one on the roof, you will have to have a chat with your landlord about it, and maybe the promise of sharing some fresh sweet honey will help smooth things over.

Another option is to see if your building has a communal garden or green space or if there are plans to make one. More and more people are looking to grow their food, flowers, or just to get outdoors for a few hours at a time. If you are living in a place with a shared garden, it would be worth it to talk to the other tenants and see if they are open to the idea of having a hive. Sell them on the idea by telling them all about the benefits bees will have on their garden as well as the general environment. Again, bribing with the promise of honey can never go wrong.

No matter where you live, it is a good idea to talk to your neighbors before getting a hive since bees can be intimidating for those unfamiliar with them. It is important to prepare your neighbors for what they can expect living next to a hive and to create a good rapport with them so that if they have questions or concerns, they can talk to you directly.

While you can never guarantee that the bees will not be pests, hopefully, it will just be the odd bee sighting here and there. But if your hive swarms, a large mass of flying bees can be terrifying. Establishing a good relationship with your neighbors before getting a hive means having friendly neighbors who are more likely to talk to you calmly rather than doing something drastic. Make sure you leave a clear channel of communication open with them, whether that be through giving them your email or phone number, so that they feel comfortable contacting you with their queries.

Gear

After figuring out the legalities and logistics of where to house your hive, your next step is figuring out all the gear you will need. Unfortunately, many of these items are very specific to beekeeping, so you cannot reuse items you have found lying around your house. This is where the cost of beekeeping comes in, and for many potential beekeepers, it is what keeps them from engaging with the hobby.

To try and prevent that from happening to you, we are going to go over what is considered the absolute essentials for a beekeeper. These essentials are the hives, the hive frames, bee feeders, a hive tool, a beekeeping suit, and a smoker. Luckily, these things are not ridiculously expensive, so you are not looking at spending tens of thousands of dollars. The price, of course, will vary depending on where you live and how many hives you want to start with. All prices (in US dollars) given in this section are based on what we could find on Amazon at the time of writing.

Hives are the fundamental part of beekeeping since that is where you keep the bees, and it is recommended that you start with at least two. That way, you can compare what is happening in one to the other and potentially use one to help bolster the other if one starts to struggle.

There are two different approaches to buying hives. The first approach is to buy a starter hive, which, while more expensive, comes with the bees and the hive. These kits often include most of the gear you will need, which is where the higher cost comes from. The other option is to buy a bee kit, which only includes a single queen and her worker drones. If you go with this option, you will have to get separate empty hive boxes, which currently cost around $150 for each hive.

When choosing your hive box, there are different style options as well. The three main kinds of hive boxes are the horizontal top bar, the Langstroth, and the Warré box types.

The horizontal top bar design is the oldest hive design in the world, featuring bars laid out across the top of the cradle-looking hive. That enables the bees to build comb down from the frame. Because of this, you do not need to use four-sided frames or a hive foundation. The advantage of this kind of box is the simple and easy access to the combs, which are easy to then remove without any heavy lifting. Though it does require frequent monitoring, those inspections are usually quick and easy.

Both the Langstroth and Warré are vertical boxes, though. As the horizontal top bar uses bars, the Warré uses frames instead. These kinds of hives are good for those searching for low-cost and low-maintenance hives that are lightweight and do not require frequent inspections.

The Langstroth is the most common now, as it has removable frames built into stacked boxes that the honeycomb is built onto. Each style of box has different pros and cons, so deciding on a style is largely a personal choice. If you have the time, tools, and expertise, you can always build a hive box yourself. This allows for even more personalization, but the overall expense will depend on the price of the materials.

The frames are the rectangular pieces that fit inside your hive where the bees make the honey. They can either be bought with your hive or separately, and you can also get them with or without foundations (an underlying base for the honeycomb to be built on) and are around $50. You could also make them yourself, but since they need to be very precise to fit into the hive, you should only do so if you are confident in your carpentry skills.

Bee feeders are used so that your hive does not decide to move when the natural food sources grow scarce. This means that having these is an absolute necessity for a good part of the year when you need to feed your bees sugar water. Feeders are wooden frames with grates over the water to prevent drowning. They are placed near or inside the hive. Placing them within the hive ensures that the food goes to your bees and not to other animals.

The hive tool is pretty cheap (around $10), but it is invaluable for what it does for your hive. Bees line their hives with a substance called propolis, which is essentially a very strong glue that holds everything together. This includes holding the lid of the hive down, as well as holding the frames in place. The hive tool is used to lever the top of the hive off, as well as to scrape along the sides and loosen the frames so you can collect the honey.

After securing the safety and comfort of your bees, you need to make sure that you are keeping yourself safe too. That means getting a beekeeping suit which will help keep you from getting stung. Honeybees are not naturally aggressive, but they do pick up on anxiety, and as a new beekeeper, you are likely to feel pretty anxious.

A good bee suit will cover your entire body and has gatherings around the ankles, wrists, and neck to prevent bees from getting inside. They then have a hat with netting around it to protect your head. You will need to combine the suit with decent and sturdy gloves and boots (tall rubber boots are best).

The best bee suits are ventilated, and though they do cost more than the non-ventilated ones, it is worth it to prevent overheating and sweating. A full suit can cost anywhere between $100 to $200. To save a little money, you could get a beekeeping jacket, which runs around $70. This one only covers the upper body but can be combined with some sturdy overalls, gloves, and boots to give you full-body protection.

The last of the essential items is a smoker ($20-$30), though the use of such devices is somewhat controversial within the community. Smoke introduced into the hive makes the bees docile and masks any pheromones they emit, meaning that they are less likely to sting you as you collect the honey. However, the smoke itself does not sedate them; instead, it puts them into a heightened state of anxiety, which results in them gorging on the honey. They are then calm because they are stuffed full.

Some critics of smokers think that using them stresses the bees out too much to make them worth using. They also say that overuse of them can reduce honey production. Using a smoker will need to be a decision you make after fully researching both sides.

The items mentioned so far are the essential ones. But, of course, there are also other pieces of gear that, while not technically essential, are helpful for first-time beekeepers to have. These include a queen catcher, queen marker, queen excluder, hive scraper, and brushes.

The queen is the most important bee in your hive because there would be no colony working to make honey without her. A queen catcher is a small plastic hair-clip-looking thing though it has rounded edges and forms a container. These are used when you need to separate the queen from the rest of the colony and are generally inexpensive, with a pack of five costing around $10.

But if you need to separate the queen, first you need to know where she is. A queen marker, less than $20, uses a non-toxic bright ink to mark her back legs, making it easy to spot her in the crowd. This is doubly important in case she dies—if you do not replace her soon enough, your entire colony will die as well.

A queen excluder is a metal grate with openings small enough for the worker bees to fit through, but not the queen as she is larger than they are. This grate allows you to control where the queen lays her eggs, which makes it unpopular for those who want to follow a more natural method of beekeeping.

A hive scraper is used to clean the sides and top of the hive. Since these spots need to be cleaned regularly, the scraper can make that job much faster. Additionally, another useful tool is a brush to knock away the bees from the frames you pull. Most often, you can shake the bees off, but a brush can be used to remove the stubborn ones who cling to the frame.

You can get these items, plus others, individually, but they are also often sold together in kits. You will need to do some shopping around to figure out which ones are best and include the tools you want, but it could be a good way to get multiple things together for a lower price. (Small kits start around $430.)

Buying Your Bees

Alright, so now that you have joined a club and bought all your gear, you have built yourself a strong foundation. It is on that foundation that the basis of your experience should rest, and you must have that base before moving onto the next step—setting up the stars of the show and buying your bees!

Buying bees does require some preplanning, but it does not need to be a complicated process. There are three different methods of acquiring your bees; buy packages of bees, buy a nucleus (nuc meaning already established) colony, or catch a swarm of wild bees.

The most common method for first-time beekeepers is to buy a package of bees. These are generally shipped countrywide in screened containers that hold around 10,000 bees and the queen, who is separated from the rest by a piece of sugar candy that the workers can eat through to get to her. You should double-check with the company you are buying from to see if the queen is marked because that will help you identify and place her where you need once the bees arrive.

There are several online stores where you can buy bee packages within the United States, Canada, and other countries. These packages will be sent to you through the mail with special carriers or through the normal mail—you will need to check which shipping method the company uses. It is also probably a good idea to double-check with your local post office to see what their policy is regarding live insects. They may require you to come to the brick and mortar store to pick them up rather than being delivered to your home.

Nuc colonies are more expensive than bee packages, but some advantages go along with the higher cost. With a bee package, you are only getting the bees themselves, but with a nuc, you are getting a few frames of premade honeycomb that were produced over the winter. This means that your new bees will be able to start producing honey much faster, and the queen is already laying eggs. If they have already survived a winter, you know that they are hardy bees. These kinds of colonies can be installed right into the hive and are most suited to the Langstroth style of box.

For both of these options, it is possible to buy the bees from a local merchant rather than from an online source. The advantage of doing so is that if the bees are local, they are already familiar with the area and therefore are better adapted to your unique conditions. When you purchase locally, you can talk to the seller about any difficulties they have had with the bees in the past and what kind of conditions the bees do not like. It can also be an extra guarantee that the species of bee is one suited to your environment.

If you want local bees but want to save a few bucks, you can always try to catch wild bees. During the spring, many existing colonies have grown too large, and they swarm to create a new hive. If you are on the lookout for such swarms, you can catch the queen bee and set her up in your hive—the worker bees will follow their queen. Swarms can look scary, but most of the time, they are very docile as they do not have an active hive to protect, and they have stuffed themselves with honey to sustain the long flights.

This method is better for more experienced beekeepers, so if you are working on your own, you may want to reconsider this method. Joining a beekeeping club can give you the advantage of gaining a mentor or other experienced apiarists who are willing to assist you with catching wild bees. However, there are other things to consider beyond just the cost and physical act of catching the bees. Because they are wild, you will not be able to know anything about their health, what their temperament is like, or any potential genetic issues from the queen.

What to Look for in Your Bees

Now that you know the different ways of getting your bees, we should discuss what you need to look for in them.

First, you will need to decide what breed of bee you want to have. Yes, there are multiple breeds of bees available—it is not just 'honeybee' one and done. The most popular breeds are Italian, Carniolan, Caucasian, Buckfast, and Russian, and each has its good and poor qualities. However, all honeybee breeds can crossbreed, so do not stress about the breed too much. Most packages of bees are a mix of different breeds to help create stronger genetics.

When selecting bees, you are looking at three elements: temperament, their productivity, and how well they will do in your climate.

Italian bees are suitable for many different kinds of environments, but they do not do well in tropical climates. They have a fairly mild temperament, are not prone to swarming, and make white caps on honeycomb, all of which have made them a popular bee for beginner apiarists. However, they are prone to starvation in the winter due to large colony sizes, so if you live in an area with long winters, you will need to make sure you supplement their diet quite a bit.

The Carniolan bees originally come from Yugoslavia and are darker in color than the Italian bee. Carni bees are calm, gentle, and easy to manage, making them a good choice to introduce to more aggressive bees, though they are likely to swarm. They prefer cool, damp climates and will more likely search for food on slightly rainy days.

If your hive has access to lots of natural nectar, a colony of Carni bees will grow quickly, though they are good at self-regulating their numbers. This means that more will be born and work if there is a lot of food, or a hive will be smaller with a lack of food. This equals a smaller colony during the winter, making starvation less of a concern.

Caucasian bees came from near the Caspian Sea and so are better suited for colder climates. They have longer tongues than the other bees, making it easier for them to get nectar from deeper blossoms. Their primary advantage is their calm nature, with many beekeepers considering them the gentlest of all honeybees. Their disadvantage is that they create a lot of propolis, that sticky glue-like substance inside the hive, making inspections take more time and effort. They are also not a common breed of a bee to have, so finding some will be very difficult.

Both the Buckfast and Russian bee are specialty bees bred from several other breeds. Buckfast bees are not as popular as they once were since the offspring of a naturally occurring queen became very aggressive. This trait was negative enough that it overshadowed their ability to produce lots of honey. As a result, in modern beekeeping circles, it is very hard to find Buckfasts for sale.

The Russian honeybee is well suited for cold climates and is somewhat disease resistant, but no bee breed is truly immune to mites and pests. Like with the Carniolan bee, they have smaller numbers in the colony during the winter, but it takes them a little longer in the spring to return to higher numbers. But once they do, watch out! Their population can grow incredibly fast, which means that they are very prone to swarming, so you will need to watch them closely.

But these are just general guidelines—most often, the bees you buy are going to be a mix of different breeds. And this is a good thing! Just look at all the problems any kind of purebred animal has; genetic diversity is a good thing as it makes your bees stronger, more productive, and hardy.

No matter what kind of breed you think is best, there are few tips to make it work the best for you. Firstly, you should ask local beekeepers what kinds of bees they have. You may find that most work with only one particular breed, which likely means that those kinds of bees are best for your local environment. If the local beekeepers are willing, also consider buying your bees from them, because as we stated before, that allows you to know that your bees will thrive in your area.

Secondly, no matter who you buy your bees from, make sure you order early in the year. The later you wait, the fewer options you will have to pick from. As well, many bee companies are slammed during the spring months, so preordering will make sure that your order is scheduled, and you will not miss out on getting your bees.

Depending on the room you have for bees, you may want to consider setting up multiple hives. Having several hives can help you monitor your bees better, as you have something to compare their activity and health to. Additionally, if one hive starts to falter, you can borrow from the other one to boost it back up. If you do not have space, time, or money for more than one hive, ask your mentor if they are willing to use one of their hives to boost yours in times of need.

Final Considerations—Money and Time

Beekeeping is neither a super expensive nor a cheap hobby. Before you commit to creating this change in your life, you need to make sure you have realistic expectations for the time and financial commitment to which you are agreeing. It would not be fair to either you or the bees if you got partway through the process and realized that you could not keep up with the demands. Abandoned hives can end up doing more damage than good, so that is what we want to avoid at all costs.

We did give an estimate of costs with some items of gear, but in this section, we will cover all items more fully. We will also discuss probable events that can occur in your first year as a beekeeper that can create unexpected costs.

The first thing to do is find out the costs of membership to your local beekeeping club. This will vary by location as well as likely by the number of hives you have—the higher number of hives, the higher your membership costs. Most memberships will operate yearly and can range from anywhere from $40 to $500. On top of membership fees, check out how much the club charges for the courses and workshops they run, and build that price into your budget.

After getting a sufficient amount of training and education on the subject, you will want to buy your gear. Here is a breakdown of ranges for it all:

- Hive boxes (depending on size and style): $150 to $300

- Accessory equipment, such as your bee suit, smoker, hive tool, etc.: $100 to $300

- Bees

  o Package of about 10,000 workers, drones, and the queen bee: $100 to $135

  o Nuc colony with premade honeycomb: $125 to $175

  o Catching a swarm: only the cost of treating potential stings

- 10 pounds of sugar from Costco to make supplemental food: $8

Added to these are the unexpected costs. You should have some money set aside as emergency funds, though one thing you can plan for is not so unknown. Mites are a common problem for all and can badly affect novices and seasoned beekeepers alike. Mites can cause a lot of damage to your hive and can spread to others, so you need to monitor and plan for them. Mite treatments cost in the range of $20 to $200, depending on how badly they have spread. So, the earlier you catch them, the better.

So now that we have looked at the financial commitment, the time needed for beekeeping needs to be shown as well. In general, expect to spend 15 to 30 hours caring for your bees in the first year.

After buying all your equipment, you have to set it up. Hives can be made of several pieces that need to be put together Ikea-style. The time needed for that setup will depend on how comfortable you are with assembly. Once the box is set up, placed in the optimal location (covered in Chapter 2), and the bees introduced, next comes the regular work.

The first thing to remember is that beekeeping is seasonal, meaning that certain times of the year will be busier than others. Also, you will likely need to spend more time in the first year on your hives until you are more familiar and comfortable with the process.

Spring is the busiest time of year and is when you will introduce your bees to their new hive. As they settle in and get used to their new home, you will have to feed them for the first few weeks. As the weather warms and they become more active, you will spend less time feeding them unless you are having a poor year for flowers. From here on, you will need to inspect the hive for disease, damage, or signs of swarming once every week. Mostly this can be done on your schedule, but certain situations need precise timing, such as introducing a new queen to the colony.

The weekly inspections continue through the summer. Once the weather starts to cool, and the leaves change color, you know the fun is about to begin. Fall is when your hard work gets rewarded with fresh honey! The time needed for honey extraction will depend on the method that you use (Chapter 6).

After enough honey has been removed, you need to prepare your hives for winter. First, complete a thorough inspection to make sure the queen is strong, the colony is healthy, and no pests have got in or can get in. There needs to be enough honey stored within the hive box to feed your bees as it is their main source of food during the winter—in colder parts of the world, bees will eat up to 90 pounds of honey in one winter (Nickson, 2019a). If there is not enough honey for them, you need to make sure you give them sugar water regularly.

You will want to close most exits to the hive and place an entrance reducer over the few holes left open at the top. The entrance reducer does exactly what it sounds like—it reduces the size of the entrance. This controls airflow and the temperature, and it reduces the chance of other bees or animals getting into the hive and stealing honey.

By mid-October, the hive needs to be wrapped with insulation to keep the bees warm. It is best to use foam insulation. As the bees move around, they generate heat, and you do not want the air inside the hive to get overheated. Wrap foam around the sides and ceiling, making sure to do so in a manner that will not trap water. The entrance hole at the top of the box needs to be left open as you need proper ventilation to prevent mold growth.

* * *

We just covered some of the basics a budding beekeeper needs to know. If any single part of this chapter can be stressed, it is the usefulness and benefits of joining a local apiary club. There you can find courses to supplement what you learn from this book, friends who are interested in the same things you are, and a mentor to help guide you.

With a mentor, some new friends, the right gear, and the desire, money, and time to get after it—nothing is stopping you now!

You can Google these terms to learn more:

- Canadian Association of Professional Apiarists
- American Beekeeping Federation
- Canadian Honey Council
- Bee Culture
- DIY Hive Box Designs
- Top 10 Best Beekeeping Starter Kits

Chapter 2:

Location Is **EVERYTHING!**

Just as in real estate, the key to bee-ing a good beekeeper is location, location, location.

The previous chapter is a wonderful introduction to the hobby and what to expect, but in this chapter, we will get into the meat of it some more. Figuring out where your hives will go is key and should be the very FIRST thing that you plan out. You have to not only ensure that your yard is a conducive environment to begin with, but you also have to choose the perfect spot for the bees within your yard as they may act differently depending on what interacts with their hive.

So, what should you do if you do not have a lot of space in your backyard or if it is not ideal for bees? Or even worse, what if you do not have a backyard at all?

Well, no need to despair because we will cover that in this chapter as well. After all, if the mountain won't come to Muhammad…

You may not have the proper space or location to house a hive, but that does not mean that you give up on the idea. There could be someone close by who has a large property and would be willing to play host as long as you agree to do the work and cover the cost of upkeep. That way, they see the benefits of having bees on their property, and you still get a brand new hobby to fill your time.

One of the first considerations is food, as bees have to travel to find the nectar they need to feed the colony. While they can travel up to five miles to do so, they will much prefer staying closer to home. So, you will need to consider their food sources and whether or not you can provide that for them or if they have to go elsewhere. And if they travel to find nectar, you will need to be aware of their flight paths—if they regularly fly over where children play in the yard, you may need to consider moving your hive. It is also important to check with your closest neighbors; if one of them is deathly allergic to bees, you should put your hive as far from their property as you can.

Once you have determined the location of your hive, it is time to look at all the other factors as well. This includes the type and intensity of sun they should have, the amount of space between the hives, and how high they should be off the ground. You also need to think about how to access your hives and how to position them so the entrance is easy to get at without being in the way of foot traffic.

Along with all these considerations, you also need to think about your bees' ease of access to water and food. If you do not provide somewhere where they can get nectar and water, they will go to the next best source, even if that is your grouchy neighbor's pool and yard.

You may be feeling a little overwhelmed right now, thinking that there is just too much to consider! Stop. Take a deep breath. It is going to be alright—we are going to walk through it all together. By the end of this chapter, you will know how to pick the perfect location and what you can do to make it optimal for bees to thrive for ultimate honey production.

The First Steps

Now that you are a card-carrying member of a beekeeping club, you are building up the knowledge to get you through your first year, and you have bought and assembled your hives and gear. It is time for the next step. Before you buy your bees, you need to create the perfect spot for your hive—it is vital to do this before your bees arrive because it is incredibly difficult and backbreaking work to do afterward.

This means creating your yard into the perfect environment for bees and choosing the right spot for the hives.

First, wherever you place them needs to be perfectly level side to side, with the front of the hive box just slightly (one inch or less) lower than the back. You put hives this way because bees build comb perpendicular to the ground. If you are on a slant, that means the comb will be as well, making it harder to work with. Also, uneven leveling makes it easier for your hive to tip over.

Have the front lower than the back. This way, if any rainwater ends up in the hive, it drains out of the box rather than staying inside. Standing water in the hive creates a drowning hazard for your bees and promotes mold growth, which can be fatal for the little guys.

To help the water drain even more and to get a perfect balance, you may want to consider getting a hive stand. When the box is off the ground, it prevents water from leaching from the soil, adds an extra boost of gravity to drain water out of the hive, and makes ventilation and temperature regulation a little easier. It can also save your back from bending and lifting the heavy frames from lower on the ground.

Certain kinds of pests and parasites thrive in damp ground, so keeping the box above ground makes it harder for pests and any predators to get in. There is a huge range in type and threat level of predators, which include bears, skunks, mice, raccoons, and wax moths. (We'll cover more on dealing with predators in Chapter 5.) There are several options for store-bought stands, but a do-it-yourself solution should be fairly simple as well.

Speaking of ventilation, make sure that wherever you put your hive, it has good airflow. You do not want it located somewhere where the air stills and gets heavy or at the top of a hill where it is open to wind and the full fury of summer and winter storms.

Having a windbreak (either a fence or hedge) is also very useful. Some areas get very little wind, while others have such strong wind that it is part of the city's identity. If you live somewhere where there are strong winds, consider placing your beehive in a location that protects the bees from strong winds, especially as they enter and exit the hive. A large fence or shrubbery will act as a windbreak as well as having the benefit of making the bees fly high above people's heads.

It is best to have the entrance of your hive facing southeast. The early morning rays will wake your bees, getting them foraging earlier in the day and therefore productive for longer. But you do not have to have the hive in direct sunlight—dappled light, such as the limited amount through the leaves of a tree, is best because too much sunshine means that the bees have to spend a lot of time regulating the temperature of the hive instead of making honey. At the same time, do not put them in the deep shade either, as that can help dampness grow and make your bees listless, which slows down honey production as well.

Speaking of honey, the final thing you need to be thinking about when you place your hives is the ease of access when you are ready to harvest the honey. You need to make sure it is in a spot that is not too far from where you want to process the honey since it would suck to be lugging dozens of pounds of honey in the hot sun as you go up and down hills or over obstacles.

These tips are great to help you put and make your bees productive, but you need to consider their quality of life as well. Bees forage for food, meaning that they fly around and collect pollen from flowers and plants, even those that do not produce nectar. This means that they often collect pollen from plants that may have been sprayed with pesticides, which can harm them in the long run. You do not need to have a lot of flowers or plants in your yard to provide food, but it can be a way to help make sure that they are staying safe, especially if you live near places that use a lot of sprays.

Additionally, you also need a good source of water for them; as a good rule of thumb, that water needs to be less than 50 feet away from the hive. We will talk more about water needs later in the chapter.

If you are in an urban area without a backyard, there are still options for you as a beekeeper. As we said in Chapter 1, placing a hive in a communal garden can be a great idea, both for the gardeners and you. If you go with that method, keep the same information in mind when you decide where to put the hive.

The other option we mentioned was to make a rooftop hive, which has many of the same considerations. However, there is more to consider about staying safe on a roof.

Firstly, if the only way to get to your roof is through a fire escape, climbing a ladder, or going through a rooftop hatch, it is not a good location for a hive. Trying to travel in these methods while decked out in a full bee suit or when removing heavy loads of honey is dangerous. If that is your situation, you need to have an alternative location for your hives.

If you can get to the roof safely, do not put your hives near the edges of it. This seems like pretty standard information since going near the sides of a tall building is never a good idea. Also, make sure you tie your hives down with strong straps or cords. Winds get much stronger on top of buildings, and it would not be good for a strong wind to come along and rip your hives apart or blow them off the side. Forget a piano falling on your head! We think a box full of bees would be way worse.

But what should you do if you do not have that kind of setup in a yard, or you can't have a rooftop or communal garden hive? Can you still be a beekeeper?

The answer is yes—if you are willing to have a commute. If you know someone in the area who has a larger yard, or a farmer not too far from you, talk to them about hosting a hive. If they are willing, you can set up your hive on their property and come out and check it regularly, just as you would if it was in your yard. That way, they get the benefits of bees, but little to none of the work required.

If all else fails, look into wooded or forested spaces around. You need to check with the city and state (or province) to see what rules are in place about beekeeping on public land, but other than that, the same principles stated above apply. Another consideration with forested areas is the increased likelihood of more significant predators, like skunks, who love to eat bees as a tasty treat. A hive stand is a really good idea to help keep the hive away from critters like that, to get it further away from their paws.

Now that you have the physical location figured out, you still need to keep in mind a good number of things. Ensuring that you are working with the best possible situation will keep your bees happy and productive, meaning that you can manage a successful hive.

Sun

When you drive through the country and see hives out in the fields, they are often in open areas without any shade. It can be easy to think that means that direct sun is the best for bees, but it is not necessarily true. Those hives are often placed in locations with full sun because there is no other option—many fields are wide open with no trees or natural objects to provide a break from the sun.

When observing wild colonies of bees, they tend to create their hives in shady areas, which makes some wonder if that is the better way to go. So, which is it; sun or shade?

We said earlier that south-facing, dappled sunlight is best for bees, but you can make a different choice. There are pros and cons for both full sun and shade, so we will do our best to outline them fully here.

Full Sun

As we said earlier in the chapter, direct sun and southward-facing hives benefit from warming up faster. Bees rise with the sun, so during those long summer days, they will have an earlier start and longer working days if they are in direct sunlight. Having the sun on the hive all day can also help keep any water that found its way into the hive box from sitting there too long.

There are also seasonal advantages to direct sun as well. During the winter, direct sun can have the benefit of keeping your hive warmer. It may feel cold outside, but with the sun warming the insulated foam around the hive, the bees stay nice and toasty. That helps you stay secure in the knowledge that your little friends are not going to freeze. And when winter gives way to spring, the warmth from the sun reaches the hive much faster, meaning that the bees can get out and about earlier in the year than those in shaded hives.

However, there are downsides to full sun as well. The first is the temperature inside the hive. Bees, just like us, do not like getting too hot—they like to stay at a nice 95 degrees Fahrenheit. To stay cool, they gather water and spread it in a thin layer over the brood comb (where the queen lays the eggs). Once the water is spread out, they stand on the comb and vibrate their wings at high speeds resulting in evaporite cooling. Essentially, they vibrate their wings so fast that air currents are created, traveling over the water layer and evaporating, just like air conditioning.

The hotter it gets inside the hive, the harder the bees have to work to create the evaporite cooling, so there are fewer bees out collecting pollen and making honey. Additionally, the more bodies there are in the hive at any one time also increases the heat since they are all moving around so much and generating their own heat. A sign of a hot hive is when the bees hang out in clumps outside of the hive—called bearding.

If the hive gets unbearably hot, the colony can swarm and try to find somewhere cooler to live. In extreme cases, when it is too hot, and the hive is not properly ventilated, the wax bombs can melt and drown the bees. To avoid this, if you have to place your hive in direct sun, make sure that there is a lot of water nearby and that the bottom of the hive is a screen, as that promotes airflow.

Shade

If you were to base your hive location on the behavior of wild bees, you would want to place your hive in an area of deep shade. Most wild bees like to nest in shaded areas that are still close to open areas, such as at the edge of a forest. This protects them from the worst of the sun in summer and gives them clear flight paths for foraging.

Some will say that leaving the hive in direct sun benefits the beekeeper more than it does the bees since there are risks associated with direct sunlight, as we stated above. However, by staying in the shade, the bees do not have such a high risk of overheating, though they will have less productive hours in the day.

There are downsides to too much shade, aside from less honey production. If you live in an environment that gets a lot of rain or has a humid and cool climate, placing your hive in the shade may make it harder for it to stay dry. Additionally, some beekeepers feel that shade encourages hive beetle infestations, though others strongly disagree with that idea.

So, What's Best?

Like so many parts of being a beekeeper, deciding how much sun or shade is a personal decision. You need to weigh the pros and cons as well as consider your environment.

If you can, it is generally considered the best approach to place your hive where there is a good mix of sun and shade. Ideally, you want your bees to be woken by the early sunlight, have a place to stay cool during the hottest part of the day, and then have some extra sunlight in the evening. That way, you (hopefully) get the benefits of both without straying into the negatives of either.

If you have a nice deciduous tree, putting your hive in its shadow is a great idea—that way, the hive gets shade in the summer, and the winter sun can warm it once the leaves have fallen.

Also, you get the benefit from working in the shade when wearing a bee suit, which can get very hot quickly. Even a little extra shade could be a godsend on a hot summer day.

Space

Another thing you need to consider when setting up your hive is the issue of space—space for yourself, your neighbors, and the bees if you are going with multiple hives.

The first concern of space is safety; are your bees somewhere where they and others will be safe? You will want to keep the hive away from areas of heavy foot traffic or frequent use. If bees have to fly through such areas, there is a greater chance of someone being stung or your bees getting squashed.

It is also a good idea to have the entrance of the hive facing toward a tall barrier, which can help direct the bees to go where you want. It also acts as another safety measure, as it can keep them out of easy visibility of your neighbors, friends, or family that are nervous around bees. Honeybees are gentle little creatures, but they pick up on anxiety easily, and they will notice if someone gets nervous just by looking at them. This could result in them becoming agitated and defensive, increasing the likelihood of stinging, so it's better to keep them somewhere out of sight. After all, out of sight, out of mind, right?

When thinking about space with multiple hives, you need to make sure that you are giving yourself good ease of access. You will need to complete regular inspections, maintenance, and eventually harvest honey, so do not put your hives so close together that you can barely move around between them. At the same time, do not put them so far apart that you have extra walking to do to get to each. When placing multiple hives, it is a good rule of thumb to have between two and three feet of open space around each hive.

Aside from making it easier on yourself, having hives too close together can create problems for the bees as well. When placed too near each other, bees tend to drift (go back into a hive that is not their own).

Bees mostly navigate by the sense of smell, and each colony has a unique smell based on the scent of their queen. They do not rely just on scent, but landmarks as well. So, if there is not enough variety between hive boxes for them to focus on, they can end up going into the wrong home. This is a problem because the established colony will view them as intruders and will often kill the new bee. To avoid drift, place your hives far enough away from each other so that the bees can distinguish between them. Painting each hive a different color will also help create unique markers on the landscape.

Having hives close together can also make the spread of pests and diseases easier. Another concern is having the bees of one colony pick up the distressed pheromones of another, making them unnecessarily agitated and aggressive. Because of all these reasons, placing hives two to three feet apart is the best method.

Height

Earlier in the chapter, we talked a little about the uses and benefits of hive stands, but we should probably look a little more into the height of hives.

As we said before, stands are a great way to get some extra height for your hive, which prevents water buildup, promotes water drainage, and makes it harder for predators to access the hive. Additionally, they are really helpful in saving your back—a stand makes the hive the right height for you to be reaching and lifting things from with as little strain as possible.

Height is very important to get just right—too low, and it makes it easier for predators and pests to bother the hive, too high, and you cannot get into them easily or safely. Ideally, the base of the hive should sit around 18 inches off the ground, though some adjustment will need to be done based on your height.

Water

Like all living things, bees need water to drink and cool themselves down (evaporative cooling). They will get that water anywhere they can, even if it comes from a bad source. And unfortunately, in most urban centers, the available water is going to be filled with pollutants. You also need to consider the potential of harm. If the nearest water source is a neighbor's pool, the bees will drink from it—and then you are worried about them drinking the cleaning chemicals, not just stinging people or getting swatted.

If you live in a more rural setting, you might think that you are alright, but even then, water contaminants get into everything. Rivers and ponds could be filled with the run-off of pesticides and refuse from who-knows-where or who-knows-what. It is better to make a water source for them yourself.

There are so many possible ways to provide water for your bees that we won't bother listing them. It can be as simple as a bucket refilled with fresh water every few days or as complicated as a water fountain feature. It is your choice, though whatever you do, that water should be less than 50 feet away from the hives and needs to include a shallow place for the bees to stand and collect the water since they can't swim.

* * *

As you have now seen, the location of the hives is everything, and there is a lot to consider when making that decision.

You need to make sure that you set up your hives for success and set up a good workspace for yourself. This means setting your hive on level, solid ground, deciding how much sun and shade they are going to get, and providing a nearby water source for them.

Any hive should be off the ground by at least 18 inches to prevent the growth of mold and dampness, as well as to protect from nosy predators. Having the hives off the ground can also save you from future backache. If you have multiple hives, there should be between two and three feet of distance between them.

You should face the entrances in the opposite direction of foot traffic. The bees won't mind, and this helps to prevent the bees from perceiving people and pets that walk in front of the hive entrances as potential threats, making it far less likely for anyone to be stung. You can place fences around your hive or your entire yard as another great way to direct the flight path of your bees away from neighbors or high-traffic areas.

And with these considerations, make sure that you are setting up your backyard hive in a way that works for you as well. Do not put the hive in a spot that is difficult or dangerous to get to or that would require lots of extra walking while carrying heavy containers of honey.

This is a new hobby, and it should be a fun one. Poor location planning can quickly remove the joy from the work, which is the last thing that we want. So, when planning for your bees, remember the title of this chapter; location is EVERYTHING!

You can Google these terms to learn more:

- DIY Hive Stands
- DIY Bee Fountains

Chapter 3:

Preparing the Hive and Your

Chosen Location

Now that you know everything about picking the proper location for your hive, it is time to look at the practical guide of setting one up. While the last chapter covered the whys behind choosing your hive location, this one will look at the how of setting up. This will need to be done before your bees arrive because, as we mentioned in the previous chapter, trying to change things once the bees are in the hive and working away is much more difficult.

So, what does hive setup require? Firstly we will cover some of the necessary time constraints and commitments, such as when to order your bees and making sure that you have enough time between ordering and arrival to set up your chosen location. With the amount of time in mind, we will look over the specifics of your hive, such as the pros and cons of different hive material choices and whether or not to buy a premade hive or make one yourself. We will also consider the hive's bottom board. The options are to have either a screened or solid bottom—we will outline the pros and cons of each choice so you can pick what is best for you.

The final section of the chapter will outline how to install your bees in their new home. This will cover what to do with both packaged bees and nucleus (nuc) colonies.

These are suggestions only, so feel free to play around with the different options to figure out what is best for you and your bees.

Time and Setup

After reading Chapter 2 and applying that information to your situation, it is time to move on to the more practical part of beekeeping. This is the time for you to ensure that you have everything you need—the gear, items, and space needed for your bees. At this stage, the only things that you shouldn't have yet are the bees themselves. That is because this is the stage where you need to be considering timing—timing for ordering your bees and the timing required to prepare for their arrival.

As we have said before, the duties of a beekeeper are closely linked to the seasons and time of year. You need to prepare your hives for temperature and weather changes before they happen, and that includes the initial setup. What follows is a basic rundown of the beekeeper's year and what needs to be done each month—variability exists based on your location.

The best time to start preparing for the bee year is actually near the end of the calendar year, in November. During this month, as the temperatures drop, bees become less active and are preparing for winter. Experienced beekeepers will be preparing their hives for winter, but for beginners, it is the perfect time to get into the hobby.

To properly prepare yourself for the next spring, you should attend a beekeeping course and buy your tools and gear during the winter (November to January). This includes buying hive boxes and preordering your bees to make sure that you do not deal with shortages or backorders in the spring. Ordering bees at the start of the new year is standard practice for experienced apiarists since it will help avoid lengthy delays in the productive season.

For most areas in the northern hemisphere, April means the start of spring and the bee season. But, that will be different in certain parts of the world, and with climate change, the warmer seasons are being pushed earlier. April has always been the month when the temperature consistently reaches 50+ degrees Fahrenheit, and new growth begins, providing pollen for the bees.

This means that you need to have your hives and locations properly set and prepared for that change in the weather, whenever it may come for you. Make sure that you have a good understanding of local weather and climate patterns so that you can give yourself the correct amount of time to set up. There is nothing worse than getting all excited and finding out that you needed more time than you gave yourself, and you have bees sitting and waiting for their new home.

The rest of the beekeeper's year is related more to care and upkeep, with May to August being mostly concerned with inspections for diseases, signs of swarming, or other potential pitfalls. September sees the harvesting of honey before the bees begin to slow down and prepare for winter in October. We will go into further details about these ongoing beekeeper duties in later chapters.

Setting Up the Hives

Now for the setup. The first thing you need to do after making all the location decisions is to set up your hives. But first, you need to decide on the kind of hive you are going to get and how much work you want to do in the assembly of them.

Chapter 2 gave a rundown of the three main kinds of hives; the horizontal top bar hive and the Langstroth and Warré hive boxes. Once you have decided which style you prefer, you also need to figure out if you want one preassembled or not. Likely it is much easier to buy one that is already ready to go and simply install it in your yard hassle-free. If you want to assemble the hive yourself, be sure to account for the time and tedious nature of this option.

The third option is to build your hive box. This should only be considered if you already have the tools and experience required for carpentry, as hives need to be fairly precise in their layouts.

But, if you do want to make your hive, aside from the style, the other major consideration is what kind of material to make it from.

Of course, with the hundreds of lumber options out there, you could have a hive made out of pretty much any kind of wood on the planet. But it is best to go with the more easily accessible, cheaper, and environmentally sound options.

Commonly, hive boxes are made out of pine, cypress, or cedar. Hives can be made out of more exotic woods as well, but with those, you need to consider the costs (a mahogany hive can be upwards of $1,500) and the potential downsides, such as negative health effects of the wood on bees. The only kind of wood that you definitely need to stay away from is pressure-treated boards, since having your bees' home made out of chemical-laced wood does not sound like a good idea.

Pine is the most commonly used kind of wood due to its versatility, wide availability, and lower price point. There are different grades (visual characteristics that can impact manufacture) of pine. The two that you are most likely to run into are knotty or clear.

Knotty pine is visually more rustic as it has the knots and imperfections of the tree present in the boards. Usually, these features are mainly cosmetic, but they can make working with the wood a little more difficult if you need to cut around a particularly stubborn knot. This grade is the least expensive. Clear pine just means that it is free from any defects or cosmetic imperfections, resulting in a cleaner and more straightforward aesthetic. This grade will be more expensive than knotty pine.

Something to keep in mind if working with pine is that it is not the best kind of wood for the outdoors. It weathers quickly, so you may need to protect it with outdoor bee-safe paint, varnish, or stain.

Cypress is a wood that is naturally good for beekeeping and outdoor uses since it produces a sap that naturally repels mold and insects and preserves the wood. But, it is less widely available than other kinds, with it found more easily in the southern United States than elsewhere in the northern hemisphere. This will make it more pricey, but the results will be well worth it if you can spend the money.

The second most popular choice after pine is cedar. The natural oils in cedar (the most common of which is Western Red) make it more resistant to rot, warping, and bug invasions than other woods. It also has a beautiful look and smell to it, making working with this kind of wood a real pleasure. Though nowhere near as expensive as cypress, cedar will still have a higher price than pine, which makes it less commonly used.

But just because you are making your own hive does not mean that you need to restrict yourself to wood. There is a growing emergence of reasonably priced, environmentally friendly, synthetic wood made from recycled plastics and wood fibers. The advantage of this material is its ability to last through all kinds of weather and that there is a low level of maintenance required (mainly just washing).

Bottom Board Concerns

No matter the style or level of work involved with getting your hive, you will need to decide what kind of bottom it will have, whether screened or solid. As with all options, each has different strengths and weaknesses.

The first advantage of a solid board is that it will keep your hive warmer, which is good for winter and early spring since the bees will start to be active earlier. A solid bottom is also meant to deter some pests, such as fire ants, from getting into the hive while also making it easier to treat mites as they do not have anywhere to escape.

The downside of a solid board is that you will need to clean it, and it can get very messy very quickly. Additionally, by having less airflow, a solid board could cause heat concerns for your bees if you experience very hot summers. Without the airflow offered by a screened bottom, your bees will need to work harder on staying cool.

That is probably the biggest advantage of a screened bottom—it is easier for air to flow through the hive, making temperature regulation easier. Along with this, a screen makes it easier to see what is happening inside the hive and allows mites to fall straight out of the box rather than needing to be cleaned up.

The disadvantages of a screen bottom are the flip side of their positives. In winter, more cold air will get in—meaning that more intervention will be needed to make sure the hive stays warm. And while pests can fall out the bottom of the hive, they can also climb into it through the screen.

The best way to figure out which will suit you better will be to talk to others in your club and see which kind they have the most success with. That way, you can use their experience to figure out what works for your area.

Housing the Bees

After picking and preparing the location and setting up your hive, it is time to install the bees when you have them. As we said in Chapter 1, you can buy either a package of bees or a nucleus (nuc) colony. Each will have different methods of installing into the hive, which may influence your decision of which to buy. Either way, the very first thing that you need to do with them is to see how they have dealt with shipping. Some dead bees are normal, but if there is a whole lot, you will need to be in contact with the seller.

Keep them in a cool place away from direct sunlight until you are ready to put them in the hive. If you need to keep them in their shipping packages for a while, make sure to give them some sugar water (1:1 ratio) to eat—misting the mixture from a spray bottle should work out well. Installation of the bees into the new hive needs to happen no more than 48 hours after getting them, which is why it is so important to have everything set and ready beforehand.

You can install a package of bees in two ways; either let them migrate in on their own or gently shake the bees inside. Both should be done only after the queen has been carefully removed and placed in the hive, with the candy barrier between her and the others pierced so they can eat through it faster. In either case, make sure you have a good nearby supply of sugar water for them to eat as they get settled in.

Since a nuc is its own little hive box, getting the bees into your larger hive is a little easier. First, you want to place the entrance of the nuc next to the entrance of the hive, remove the screens from each, and then wait 24 to 48 hours. The bees will circle above the hive at first, but that is their way of getting situated in the landscape—eventually, they will move into the bigger space.

If they do not migrate naturally, you will need to intervene a little more. Suited up with the proper gear and tools, open up the nuc and remove a few of the frames, which you will place into your hive. Doing this is a good idea, even if they are migrating on their own. This way, they get to keep the combs (and potentially larvae) that they have already created. It also allows you to inspect them for disease or damage.

After moving the bees into the new hive, you can leave them alone for five to nine days, aside from regular feeding of sugar water. After that initial nine days, you will need to do a more in-depth inspection to make sure that the bees are doing well and that the queen is out and about doing her duty of laying tons of eggs. If you use a package of bees, make sure that the candy has been broken through, and the queen is free to move around as you want her to.

* * *

This chapter was pretty short and sweet. The main thing to remember is proper time management. Order your bees early in the year so you do not deal with any delays, and make sure that you are 100% set up and ready before the bees arrive. Doing this will ensure that you do not deal with an overabundance of stress, frustration, or feelings of defeat.

After choosing and prepping the hive location, make sure that your next step is picking the hives themselves. There are several different options for style, as well as options to buy them premade or with assembly required. If you are ambitious, you can even build a hive from scratch. With your hive, you need to decide whether you want a solid or screened bottom, though figuring that out may require asking advice from the members of your club.

Once the location and hives are set up properly, and your bees have arrived, it is time to get them settled in their new home. How you do so will depend on whether you bought a package of bees or a nuc colony, but either way, after 24 to 48 hours, you need to do a quick check to make sure the queen and other bees are free and working.

There is quite a bit of variation and experimentation that can be done with all this information. The main thing is to give yourself the proper time needed, but other than that, make sure that you are relaxed with it all. It is a new hobby, and it is meant to be fun still. The things above are just some considerations—feel free to experiment with what works for you and your bees.

You can Google this term to learn more:

- 38 DIY Hive Plans

Chapter 4:

Look After Your Bees!

Your yard, hives, and bees are now all prepped and ready for the next step, and so are you! It is time to learn about the proper care for your bees and what that looks like.

You want to ensure that you continue to look after your bees and make adjustments to the hive as necessary while you learn, adapt, and your knowledge continues to grow. You are starting a long-term relationship with these bees, and just as you would with a romantic partner, your bees require your tender love and care to thrive.

This chapter is all about proper maintenance, showing you what you need to be aware of and what you need to focus on the most to fully succeed as a beekeeper.

The first section in this chapter is going to be a detailed overview of how to fill your hives with bees—this part will go into more depth than what was covered in Chapter 3, which mostly discussed methods. So, while it is likely there will be some repeated information, it'll be handy because this way, it can help stick in your mind more!

An important part of re-homing your bees includes the successful installment of the queen bee. If you goof up on this section, it will cause both you and the bees stress, so you want to make sure that you are doing the best method the first time. It is the queen who attracts and keeps the rest of the colony in place, so getting her settled in first is the top priority.

Once we have gone through the details of settling your queen and bees into the hive, it is time to look at what else is needed to keep them alive and happy. This includes more short-term, immediate needs, such as feeding them while they get settled in, and long-term concerns like setting up your yard for pollination.

The final part of this chapter is all about hive inspections—how often you should be checking them and what you need to be looking for as you do so. It will also cover what you will need to do when encountering any issues, such as how to repair or replace damaged parts of the hive and how to replace and clean up lost beeswax. Issues of ventilation, shade, and rain will also be discussed in this part, though briefly.

With all this knowledge, you will be well on your way to becoming a happy and successful beekeeper! If you look after your bees properly, it will save you time and heartache (and likely backache) in the future, as you will be more than prepared to deal with any potential pitfalls.

Filling the Hive

We may have mentioned it before, and we will say it again right now—filling your hive with bees is a crucial step in bee-ing a beekeeper, if not the most crucial. After all, can you even call yourself a beekeeper if you do not have any bees?

So, with it being an incredibly important step, that means that there needs to be a solid foundation of knowledge to draw on. That can be achieved through reading this chapter and additional written sources about it, but if your beekeeping club offers courses on this subject, make sure to take them! That will let you practice the installation with someone who has done it before, who can give tips or advice on the best methods.

We should first lay out the things that you are going to need for installation.

- **An Assistant**: You are about to be handling thousands of bees, so it is best to have an extra set of hands to help out. This job would be an excellent use of your mentor, as they can then help out with their experience as well as their hands.

- **The Bees and the Queen**: If you bought a marked queen, that is all the better to get started with. If she is not marked, doing so now will save you hassle in the future of trying to find her.

- **The Hives and Frames**: At this stage, your hives need to be completely set up and ready to go, no matter the style. Additionally, if you have a hive divided into different box sizes, the largest needs to be the one where the queen will lay the eggs, while the smaller ones are perfect for honey collection.

- **Tools and Gear**: This is your first opportunity to wear and use all the fancy gear (Chapter 1) that you bought! You should be outfitted in your beekeeper suit or jacket to protect yourself. If, for whatever reason, you do not yet have a suit, just wearing a veil can work, but make sure that you wear all white. (Bees' natural predators are dark in color, and so they do not see white as a threat.) Along with the protection, have your hive tool handy for any lifting or prying that may be needed as you open the package or nuc of bees.

- **An Entrance Reducer**: An entrance reducer is simply a barrier of some kind (wood, metal, etc.) that is placed in front of the hive entrance to stop the bees from leaving in large numbers. It can also act as a barrier against invading pests, so it could be something left at the hive full-time. When you are introducing the bees to the hive, the entrance reducer helps stop them from leaving while you are trying to get them set up, which is unlikely but still a possibility.

- **Food**: The bees will need some initial food to eat as they get settled into their hive. This includes having a spray bottle of sugar water to put a thin layer of the solution over the frames. For longer-term feeding, place a container (like a simple mason jar) or sugar water near the hive entrance or a pollen cake on top of the hive. Most packaged bees are fairly young and so are not yet skilled in foraging. Placing sources of food nearby keeps them fed while also getting them more familiar with their hive as a landmark.

Now that everything is organized and set up, it is time to get started! In total, the process of installing your bees takes around two hours—make sure that you have given yourself enough time to get the job done in one go.

The first step is to lightly spray your bees down with the sugar water (a 1:1 ratio for the spray water). It is important to make sure that you use a new spray bottle so that there are no leftover chemicals or harmful substances that can get on the bees. Spraying them with the sugar water gets them eating, and bees that are full are more docile and easy to handle. Additionally, the lack of an established queen or honey will make them easier to handle.

Once you have sprayed the bees down, take some time while they eat to get all the frames out of the hive so you can spray down the walls next. This gives the bees an even better boost of food, but this time they have to travel along their new hive to get it, meaning that they are familiarizing themselves with the space at the same time. Make sure to give the frames themselves a quick spray before replacing them in the hive.

The next steps are about working directly with the bees. First, if your package of bees included a sugar-water feeder (which it should have), you need to carefully remove it and brush the few bees who may be hanging on. This is your first physical interaction with your bees, so you need to be calm and relaxed, which helps them feel relaxed as well. If it has been a while since you fed them, give them a quick top-up to fill their bellies.

Once you have the sugar-water feeder removed, it is time to remove the queen from your package. Different companies will have different methods of transporting the queen. If you have one that had the queen in a separate container blocked off with sugar candy or paste, remove a little bit of the block and then put that into the part of the hive planned for the queen.

It is important to keep the queen in her container or separate cage for a few days before she is free to roam around, so you do not want to remove all of the candy or paste. Leaving it in place means that the bees have to eat their way through it, ensuring that the queen permeates the inside of the hive with the pheromones that keep the rest of the bees in place. If she gets loose too soon, you run the risk of her leaving the hive or being killed by the other bees. If you have a nuc, this is less of a concern.

When you are putting your queen into the hive, make sure that the cage is somewhere accessible so you can remove it in a few days. One trick to make this easier is to loop some wire into an opening of the container and then hook that over the edge of the hive, making it easier to find and remove when the time is right. After three days, check the container and make sure that the queen is now free.

If you have bought a package of bees that does not have the sugar candy blockage, then you will need to create that barrier. This is easily done by placing some sugar-candy material in the exit area. You can make your own kind of blockage, but you need to make sure that it is the right kind of food for your bees and also something that will not take them long to get through. To make it easier on yourself, you can just put a marshmallow into the exit as a block instead.

After getting the queen and her container into the hive, it is time for the rest of the bees to get in as well. You can do this by leaving the package or nuc close to the entrance of your hive and let them migrate naturally, but this will take more time and include some additional risks (such as them flying away or attracting pests and predators who want to eat them).

The faster method is to open up the hive, hold the package of bees over it, and gently shake until they all drop inside. If you have some bees that are stubbornly clinging to the sides, you can give them a little extra firm shake to dislodge them. Keep shaking until most of the bees are in the hive—any stragglers can be left in the package, angled toward the hive entrance, so they know where to go to rejoin their friends.

Once all the bees have successfully U-Hauled into the hive (around an hour or so), it is time to close up shop. Place some sugar water close to or on the entrance reducer, close the top lid, and place a pollen cake or other source of food on top of the hive.

After three days have gone by, remove the queen container from the hive. This is where the wire hook comes in handy—dropping the container into a hive full of active bees would not be a fun experience. Ten days after installing the hive is the time for the first in-depth inspection to make sure that your queen is laying eggs and if the comb is starting to form.

Other than the queen cage removal and inspections on days three and ten, there is very little to do for the first stretch of time as a new beekeeper. However, one significant thing that you need to do is monitor their sugar water levels to make sure that they always have enough. You should continue to give them reasonable amounts of sugar water for a month or until you can see that they are gathering enough pollen on their own.

If you are concerned about amounts, it is better to have lots available than too little. Of course, it is a personal choice whether to reduce the sugar water right away or keep it going longer, but there are some benefits for the longer feed.

If you bought a package of bees, keeping sugar water available for the bees will help promote the faster creation of the combs. The combs, made of beeswax, give the bees a place to live, work, store food, lay eggs, and raise the hatched eggs. In short, the comb is essential to the hive, and to make it, your bees need food.

Talk to your mentor and beekeeping club friends about the length of time to feed your bees sugar water. Like we just said, it is a personal choice, but they may have some good insight into what your area needs. Options for the length of time include waiting until a few frames of the comb are made, for the brood boxes to be full of comb, or just to keep feeding all-season or until the bees get bored of the sugar water. So, there are plenty of options, and leaving the sugar water available a little longer is not going to hurt your bees.

And there you go! Your bees are all set up in their new homes.

Keeping the Bees Happy

Now that you have a hive full of buzzing bees, it is vital to keep them safe, productive, and happy. There are many different ways to do this, including setting up your yard for optimal pollination (if you can), staying on top of hive inspections (and knowing what to look for), and managing any damage your hives may sustain. It is also important to keep an eye out on the comb and know how to clean up any lost comb and what to do if many have broken. And as the year travels into the hotter summer months, make sure that your hives have the proper ventilation and shade to keep your bees working on honey-making instead of air conditioning.

A Pollinating Yard

As we have mentioned, it is crucial to keep your bees fed since that will directly translate into how productively they work. Sugar water is a great first step, but getting back to natural options is even better. If you have a yard, here are some tips and tricks to make it into a pollinator's paradise. Pollinator plants will attract more than just your honeybees as well. Red plants and flowers attract hummingbirds, and fallen foliage is perfect for butterfly nests in the fall.

It may seem counterintuitive to common practice and possibly a hassle with a homeowners association, but keep your yard messy. In the spring, leave some leaves and twigs around in a dedicated messy spot. Additionally, do not get overzealous in weeding your lawn or garden— most times, the common types of weeds produce lots of pollen that your bees can collect.

If you want to plant some plants specifically for pollinators, make sure that you are choosing the right kinds. These plants need to have continuous flowers throughout summer, with single flowering plants being the best since they lack extra spaces or parts that make the pollen harder to get. For information on the best kinds of plants to grow, talk to your club and do some research into what grows well in your area.

Another way to attract more kinds of pollinators is to build or buy a bee hotel. These tiny structures are full of tubes (made of bamboo or drilled holes in a piece of wood) that are the perfect nooks and crannies for other species of bees or insects to rest. Most often, the bees that use these hotels are species that are solitary rather than hived, and they may even lay some larvae in the hotel. To make it extra hospitable, have a small container of sugar water near to the hive or scattered around the yard so that any tired bee can get a boost of energy and keep buzzing on its way.

In the fall, you can also make a bumblebee nest out of a pot, some moss, and hay. Wild bumblebee queens will seek out a safe, warm, and dry spot to hibernate over the winter, so creating a little nest for them is an extra method to keep them safe.

Use all these ideas, and you will see so many different kinds of pollinators using your yard, from bees and butterflies to hummingbirds, if you are lucky.

Hive Inspections

In this section of the chapter, we will outline the steps for the more time-consuming kind of hive inspection, which is done once at least every ten days. But, those are not the only kinds of inspections and times you should be checking on the hive. Daily visual inspections of the hive, specifically around the entrance, help to head off any issues that could become threatening if not treated right away. These kinds of issues include pest interaction or robbing activity, such as what is done by yellow jacket wasps or skunks. If not dealt with right away, these invaders could potentially kill your hive within a week, putting the danger within the free time between full hive inspections. How to watch out for these pests is covered in Chapter 5.

It is important to continue to check in with your hive and the structure within which the bees are now living, and doing these inspections will be your main duty as a beekeeper. Because this is needed for success, it is important to stay on top of daily look-throughs and do an in-depth inspection every seven to ten days. For this section, the inspection instructions will be covering box-style hives, such as the Langstroth or Warré types.

To complete an inspection, you will need to smoke or otherwise calm your bees. Then you will need to remove each box, look at it and the frames inside over for damage or danger until you reach the bottom layer of the hive. An inspection needs to happen every seven to ten days, with your first few occurring closer to the former than the latter, but not more frequently. Too many inspections will bug your bees, making them more anxious and prone to stinging.

You will want to pick your inspection days carefully, as you want to do so on a warm, dry day. Only inspect on a wet or cool day as a last resort, as that will bring extra water into the hive, which is not good. Try to time the inspection during a part of the day when most of the bees are out foraging so that you have fewer on hand to work around.

To complete an inspection, make sure that you are wearing your bee suit or jacket, use a smoker (if you so choose), and have your hive tool on hand. It is also a good idea to have a notepad and pen available to take notes of anything you see during the inspection. If you have any sugar-water feeders for the bees, now is the best time to refill them.

So, first, even before putting on your suit, make sure that you have a solid plan in mind for your inspections. The purpose of this deep-dive inspection is to make sure that your bees are healthy and doing what they are supposed to—namely, the queen laying eggs, those eggs hatching into larvae, and the worker bees taking care of the larvae, making comb, and producing honey.

When you put on your suit, make sure that it is comfortable and secure against any bees getting inside. Make sure that you pick where you stand carefully. Do not stand directly in the bees' normal flight path as they may fly out en masse to escape the disturbance you are causing. If there is any wind, try to position yourself so that it is blowing over your shoulder; otherwise, the smoke could end up in your face rather than on the bees. For the last tip of positioning, try to make sure that the sun is shining over your shoulder so that you provide shade—the hive is normally dark inside, so having a sudden bright light there could cause your bees some worry.

Once you are snug in your suit and have picked your work location correctly, get the smoker activated and pump cool smoke directly into the entrance to subdue the guard bees. Once they are smoked, open the lid of the hive just a little, and get some smoke into the hive that way. Quickly close the lid and wait for a minute or two for the smoke to start calming the rest of the bees.

Make sure not to put too much smoke into the hive. Smoking makes the bees think there is a fire, so they produce pheromones to let the rest know that they need to start gorging on the honey in preparation, which makes them lethargic. Too much smoke covers all of the communication pheromones they emit, meaning that they get scared and confused, making them more likely to sting.

Once enough time has passed with the smoke inside the hive, remove the outer cover and place it upside down on the ground, far enough away that you do not worry about stepping on it by accident. If your hive has an inner cover, pump some more smoke into that section and wait another few minutes. The next step is to gently pry the inner cover off, using the hive tools to break through the propolis or beeswax. Once the inner cover is off, place it on top of the outer cover—be careful, as this may have some bees hanging on, so do your best not to squish them. Make sure to move slowly and methodically—fast and abrupt movements are certain to agitate your bees.

The next step is to remove the top box of the hive—this is usually called the honey super since it is the part of the hive where honey is collected. You will most likely need to use your hive tool to pry the top box off the lower box where the brood lives (often called the brood box). You need to repeat these steps—smoke, wait, remove—for all the boxes all the way to the lowest box.

Once you have smoked the bottom box, you then remove the frames one at a time and carefully inspect them. It is a good idea to inspect them the same way each time so that you form a habit that will help you get through them faster.

You are looking for the queen to ensure that she is still alive and that you do not accidentally kill her during the inspection process, parasites or pests, and how many frames have been filled with a comb that is ready for honey (drawn out). When you first start out keeping bees, your hive may not have the honey super or additional boxes. Once seven out of ten frames are full of comb, you can add the additional boxes.

Looking for the queen is easier if she is marked, but still possible if she is not. The queen looks different from the other bees. She will have a long and slender abdomen that is not striped, and she will be surrounded by a circle of worker bees. If you still cannot find her that way, look for where the eggs and larvae are since she will not be very far from them.

The eggs themselves can be difficult to spot for new beekeepers since they are very small and look more like grains of rice than an egg. They will be found on the comb, one in the center of each cell. To see the eggs more easily, hold the frame up to the sky, sun shining over your shoulder and into the frame. The larvae will be easier to spot if the comb is uncapped (second stage of growth). Instead of a small grain-sized item, there will be a wriggling little brilliant white bug! There will also be larvae that are capped or sealed into the comb cells (the last stage before they become adults). Ideally, there should be a ratio of 1:2:4 for the eggs, larvae, and sealed brood since they are eggs for three days, larvae for six days, and sealed for twelve days.

Once you are done inspecting each frame, replace them in the hive box in the same order that you removed them. Once all the frames have been inspected in the bottom box, move onto the middle box and do the same, replacing the entire box once you are done. Repeat these steps for every box on your hive until there are none left on the side, then put the inner and outer covers back in place. If you have any notes, write them down now right away, so you do not forget anything.

Repeat all these actions for every hive that you have, and when you are done, leave the smoker to burn out and cool on its own. If you need to, make sure that you clean your bee suit, so it is ready for the next inspection in seven to ten days.

Dealing With Hive Damage

During your inspections, make sure that aside from checking over the bees, you are also making sure the hive boxes themselves are still in good condition. The hive is outside year-round and is subject to all the weather that mother nature throws its way. No matter what, repairs and replacements are going to be required at some point. You may even need to repair the damage that you accidentally caused when taking them apart or scraping the propolis off with the hive tool!

These kinds of interventions can include full replacements or just minor repairs. If you have a small spot of rot on one of your boxes, see if you can cut it out before buying or making a whole new hive. This could include gouging out a small spot or cutting off a larger section of an entire wall. Whatever it needs, make sure to get it done quickly—once the rot starts, you cannot stop it, and it will spread to the rest of the hive if not dealt with.

Once the rotten section has been removed, you need to replace it with some fresh wood. Make sure that you use the same kind of materials as the rest of the hive; otherwise, you will be looking at different rates of breakdown. With the new piece in place, give it a quick paint (water-based or latex) or varnish to help protect it from the elements. It is up to you whether or not the new paint or varnish matches the old, but if you go with a different color, it is best to go with a light one that will not soak up as much sun.

If you can, work repairs and replacements of regular wear and tear into your yearly schedule. During the winter, move the bees between hive boxes so that you can bring in the older hive and give it a complete makeover while not having to worry about the bees. If you do this, give the inside of the hive a quick scrub as well, going as far as to scorch it lightly with a torch—and then scrape the burned part away since this will help remove any mites or parasites that may be burrowed into the wood.

Keeping an eye out for any damage and stopping it right away is the best method of extending the life of your hive. Scheduling a full hive overhaul and repair session during the off months and rotating your bees between new and older hives is another great way to prevent any major issues during the months when you want your bees to make as much honey as possible.

Beeswax

As the season progresses and your bees stay hard at work, there is inevitably a buildup of beeswax, from which they make all the comb. This can be seen in old comb that is no longer being used, oddly shaped comb that is 'growing' in the wrong way or just located in the wrong place so that inspection is extra hard. When that is the case, you need to remove that wax buildup—but be careful. Remove too much, and the bees will start eating even more honey to rebuild them.

You can remove the beeswax with your hive tool since it is perfect for scraping away the excess. But what do you do with it once it is removed? Well, store it and build up your collection first. There are hundreds of uses for beeswax—maybe you can even include a nice sideline of beeswax products that you sell alongside the honey. If you want to do that, you will need to safely melt it and reclaim the cleaned wax.

Temperature Control

We discussed issues around keeping your hive cool and warm before (Chapter 2), but it is important information, so we will briefly go through it again and add some additional tips and tricks that are good to know.

So, as has been established, bees are pretty good at regulating their temperature, but that does not mean that you should not be doing anything to help them. The less time and effort they put into keeping cool or warm, the more time they spend making the honey!

There needs to be the proper temperature, condensation, and humidity inside the hive for your bees to be comfortable, but too much (or too little) of any of those can have nasty side effects. You need to remember this because it is easy to take ventilation or temperature control too far, making it, so that the bees then work harder on getting it back to normal. A certain level of warmth inside the hive is normal, as is a little bit of condensation dripping from it since they can use that water for their cooling systems.

So, what can and should be done then? First, to help your bees, they need good airflow through the hive. Having them off the ground helps with this, as it opens it up more, gets them away from the wet ground, and can increase airflow through the bottom if you have a screen there.

Some beekeepers will also prop open the top of the hive or make large openings for extra airflow, but too much of this is not a good thing. When left alone, bees will naturally try to close those large gaps since they like a few small entrances and exits. Besides, permanently open sources are going to cause problems, depending on the weather you are experiencing.

Instead of making permanent holes or leaving the top open, drill some holes into the outside of the hive that you can block off with a cork. That way, you can pull them out if your hive needs a little more air and close them back up if the conditions are perfect for them.

As we mentioned in Chapter 2, a mixed area of shade and sun is perfect for your bees. That way, they can be warmed up in the morning, protected from the hottest part of the day, and then be a little warmer into the evening with the last rays of light. If you have them under a deciduous tree, that has the added benefit of letting more sunlight in during the winter and acting as a barrier for rainstorms during the summer.

Speaking of winter, we should talk a bit about preparing your hive for the colder months. The first thing that you need to do is take away the top box of your hive (the honey super) and store it somewhere inside—if left out in the elements, not only is it more susceptible to damage, but they often get infested with moths which ruin the comb and honey.

During the final inspection of the season, remove any excess propolis or beeswax present so that you can wrap or insulate (they are different) your hives. There are some differing opinions about these practices, with some beekeepers seeing them as unnecessary or even bad, but doing so can be an extra way to secure the lives of your bees during the cold.

Wrapping involves taking roofing felt or tar paper and wrapping it securely around the hive, leaving the entrance and ventilation holes open. It does not do much in the way of preventing heat loss, but it sure does add a lot of heat into the hive during sunny days and protects the wood. On the other hand, insulation adds materials around the hive, such as a foam board, which allows not too much heat to be lost.

Either option is available, or you can choose neither. It is up to you. That choice comes down to preference, what your winters are like, and what advice you can get from other local beekeepers. Whatever you choose, make sure that you research the reasons for doing so enough that you feel confident in your decision.

* * *

Looking after your bees is the most important part of being a beekeeper. First, of course, you need to make sure that they are safe, happy, and productive. But that does not mean that you have to spend a significant amount of time or labor working on upkeep. By careful planning and regular inspection, you will be sure to catch any problem before it can become destructive.

Planning and preparation are crucial for all steps of bee care, from the introduction to the hive to the weekly inspections, regular maintenance, and keeping the bees comfortable as the weather changes. The amount of work necessary to maintain a positive and productive beehive is minimal so long as you stay aware and in front of any potential downfall or "bumps in the road." What might those bumps be? Read on to Chapter 5 to find out!

You can Google these terms to learn more:

- Build a Bug Hotel
- Make a Bumblebee Pot
- How to Clean Beeswax: Easy Tips for Success

Chapter 5:

Bypassing Potential Pitfalls

You are well on your way now! With your hives and bees set up and cozy, and regular inspections taking place, it is easy to get into a good routine of care and concern. But, there are always unexpected problems that you can run into that set you back if you have not properly prepared for them. And while you can't prepare for every eventuality, you can be forewarned on some of the more common pitfalls that plague beekeepers.

That is what this chapter will be going over—how to stay ahead of the curve and avoid any potential roadblocks or prevent negative events when managing your new hive. These are some of the most common mistakes that new beekeepers make, so being aware of them beforehand is your best bet in avoiding making them.

The first section of the chapter will look at issues that can spring up from your style of beekeeping rather than outside threats. The first pitfall revolves around the queen, or more accurately, the potential problems of not having a queen anymore. Becoming queenless is a huge dilemma that you want to avoid at all costs since the hive is based around her. We will look at the signs of the lack of a queen, the possible reasons why you have lost the queen, and what to do next before you lose the rest of your bees.

The next few topics in this section will cover technical mistakes you might make. These include harvesting too much or too little honey, not feeding new colonies enough, improper wear of a bee suit, and incorrect use of a smoker. A common reason these problems pop up is if you have limited knowledge of beekeeping before starting and thinking that it is enough. It is always better to be over-prepared than under-prepared, especially when it comes to your bees. Also, never be satisfied with what you already know! There are numerous opportunities to keep learning and grow your knowledge base—doing so will only serve to make you a better beekeeper.

The second section will look at outside threats and pitfalls and how to deal with them if they arise. Keeping an eye on what mother nature is doing to your hives is paramount. You need to know how to look out for condensation and dew buildup and avoid it at all costs.

Pests of all shapes and sizes can also create a headache for you, so we will also cover how to control small pests such as ants, beetles, and mites. Bigger predator concerns include, among other things, bears, raccoons, and, potentially, the most damaging of all—children. Kids may be curious about the hive, and in poking around it, accidentally destroy it or get stung for their curiosity. The final part of this chapter will look at keeping your bees and kids safe by child-proofing the area.

There are innumerable things that can go wrong or negatively affect your hive, and you can never be fully prepared for them all. But these are the most common pitfalls, and being forewarned is being prepared, right?

Beekeeper Pitfalls

As a brand new beekeeper, it is easy to make mistakes. This may happen because of your inexperience, like forgetting something important or not knowing what to begin doing. The first two are bound to happen until you get the hang of things, but the last is something that you should do your best to avoid at all costs. There is nothing worse than having something bad happen with your hive and then realizing afterward that it could have been avoided if you had had just a bit more information to draw on. This section is designed to be that extra information in your hat—something that you can draw on to avoid making massive mistakes that could hurt you or your bees.

Queenlessness

Queen bees are the most crucial part of your hive, so losing her is a quick recipe for disaster. Or, it will be unless you take the proper steps to remedy her loss as soon as possible. But losing a queen can be a subtle thing without much change to the hive—at least at first. Given enough time, the changes are very dramatic, and the problem is that when those dramatic changes start happening, it becomes harder to stop the damage that has already been done.

So, it is best to stop those changes before they happen. The changes will be slow. At first, everything will look normal, with bees flying in and out of the hive and building comb. Then the number of bees starts to fall, and they become more aggressive. On a closer look, you will see that only comb and honey are being made—there are no new eggs or larvae. Checking for both stages is important in the hive inspection since a dearth of either or both can indicate that the queen is missing or not laying as much as she should be, which is around 2,000 eggs a day.

Aside from the lack of eggs and a brood of larvae, a queenless hive can be seen in the increase of honey and pollen. The bees that had previously been caring for the baby bees will now have switched to foraging and honey production, meaning that there will be a short period of dramatically more production occurring. An excellent way to see that this is happening is by knowing what activity usually happens on which frame—if a brood frame is now being filled with nectar and honey, that is a sign that no queen is laying enough new eggs.

Another indicator of the lack of a queen can be the presence of a queen cell or cap. These cells are different from regular egg cells since they are designed for the sole purpose of growing a new queen bee. The presence of these on their own does not mean that you lack a current queen, but if they show up around the same time that there is a decline in the brood, it is a pretty good indicator that your bees are working hard on getting a new queen in place. A colony that has recently made a new queen will appear queenless for a short amount of time until she gets used to her new role.

To test whether or not you have lost your queen, put a frame of a young brood from another hive into yours. If the bees begin to build queen cells on that frame, it is pretty hard evidence that there is no queen in your hive. Your options then are to let your bees finish making their new queen or buy a new queen that you can install yourself.

But timing is critical in this—if you wait too long, the worker bees will start laying eggs themselves. Those eggs will not be fertilized and result in more drone bees (male bees that are only there to fertilize eggs). Fewer workers will be going out and foraging for food. The increase of drones also means that the bees' instincts will be to kill any new bee—trying to introduce a queen at this stage will likely result in her murder. Upon seeing the signs of laying workers (multiple eggs per cell, spotty brood pattern, lack of worker bees, and increased drones), many beekeepers will write off the hive as a total loss since it is so hard to bring back.

So, if you think you have lost your queen, do not despair. Depending on how long she has been missing, there is still hope. Worker bees do not start laying eggs until two to four weeks have passed without a queen—if you are inspecting your hives regularly, you should be able to catch this much sooner.

Inspecting the brood is your best bet for figuring out how long a queen has been missing. If you see eggs (that were not laid by worker bees), your queen has only been gone for three days, which allows you plenty of time to get a new one in place. If there are no eggs, but you still have uncapped larvae, there is a little less time, but you'll still have room to work. But if there are only capped larvae, then that is your last chance to get a new queen in place since they will be the last generation to reach adulthood.

If you still have time to manage the situation, you have two choices—either let the bees naturally create a new queen or buy one and install her in the hive. Time is a crucial consideration for both. It takes 15 days for a new queen to be raised, and then it takes another five days for her to mate and start laying fertilized eggs.

Aside from these time constraints, your bees also have a narrow time window to start the process of raising a new queen. The new queen must come from a fertilized egg so that only gives them three days to realize that they have lost the old queen and to start feeding the egg royal jelly (a secretion that is fed to all larvae for the first three days, but only the new queen after that). If you have multiple hives, you can take a brood from a colony that has a queen, which will be fertilized, and introduce that into the struggling hive.

The other option is to introduce a new queen, one that is already mated. Similar to the process of installing your hive in the first place, the queen needs to be in a separate cage that is blocked by a sugar candy barrier. It takes time for the bees to eat through the candy to get to her, giving them time to get used to her scent and pheromones. This creates a better chance of them accepting her rather than seeing her as an invader to be killed. To prevent her death, watch what happens when you introduce the cage. If it gets attacked by workers, take the cage out and try again in a few days.

However, most often, if you introduce the queen early enough, the hive will want the new queen and readily welcome her into the colony. That is why timing is critical. Carefully search for eggs and broods while completing your regular inspections, and you can avoid the problems that would come with waiting too late to get a new queen.

Honey and Food

Another common mistake that new beekeepers make that can have dramatic effects on their hive is harvesting too much honey or taking it too early. Bees need honey to help them get through winter (and other periods of low natural pollen), as it is their main source of food—they then turn that food into energy to keep themselves warm and active. So, when you are harvesting your honey (Chapter 6), make sure you are not taking too much.

The amount of honey that they need for the winter depends on numerous factors: the climate where you live, ventilation, size and shape of the hive, number of bees, how much intervention you plan for, and so much more. Because of all these factors, it can be impossible to plan the exact amount needed, so it is better to leave more honey than you think is needed rather than less.

When preparing the hive for winter, make sure that you check all frames and boxes for honey. It can be easy to assume that the lower boxes are full of honey because of their weight, but it is best to check for sure before you take away the honey super.

On average, there should be between 80 to 90 pounds (36 to 41 kilograms) of honey per hive for the winter. If you live in a warmer climate, you can do with less. A colder environment needs that amount or a little more depending on just how cold it gets. To calculate the amount of honey in a hive, figure out how much a full frame of honey is supposed to weigh, which should also have been included information when you bought the hive. Then, multiply that by how many frames are in your hive. Sorry about the math—it is a necessary evil.

For a quick tip, generally, a large frame holds eight pounds (four kilograms), and a medium-sized frame holds six pounds (three kilograms). But it is always best to double-check the specifications of your hive.

Even after knowing how much honey should be in the hive, it is important to check in periodically to make sure that the bees still have enough left. When checking on the hive in winter, choose a warmer day, and keep the inspection as brief as possible so that not too much heat is lost in the process. If you have found that there is not enough honey left for them to eat, you need to be giving them either sugar water, syrup, or more honey to eat instead. If you give them additional honey, make sure that it has come from your hives so that you can guarantee that it is disease-free.

If you need to feed your bees, make sure that you are putting the food, either honey or sugar water, inside the hives. This way, it reduces the chances of the food being robbed by other insects or pests.

Dry white sugar can also be used as a food supplement for stronger (i.e., more established) hives. With this method, make sure that the bees have access to sufficient water, as they will need that to liquefy the crystals for their use. So, it is possible to use it dry, but many beekeepers prefer either wetting the sugar a little or just going with a sugar water (1:1 ratio) or syrup (2:1 ratio) system. Since it is thicker and has more sugar, Syrup should only be used when the stored honey is low—such as when first introducing the bees or preventing them from starvation in the winter.

Whatever method you choose, it is vital to make sure that your bees have enough food to survive through lean times. If you are uncertain about how much to feed them or how often, it is better to overfeed your bees. This prevents starvation and the potential death of your colony.

Proper Suit Wearing

To quote a popular character from a show supposedly telling a story about parents meeting—suit up! Another common mistake first-time beekeepers make is to either not wear a bee suit or to wear it improperly. Bee stings hurt, and in the case of allergy, can be very dangerous. You need to wear the suit properly to prevent any harm coming to you—no matter how friendly bees are, you are messing with their home, and that will get them riled up.

Make sure that you wear a hat and veil. You will need both because stings in the neck or the face are incredibly painful, and the stingers are difficult to remove. This part can either be attached or separate from the main body of the suit, but either way, make sure that it is sealed correctly and has absolutely no open spaces that a bee can fly through. If the hat and veil are separate, it is best to get the kind that zips together.

The main body of your suit needs to be white for two reasons. The first is that bees do not have white-colored enemies, so they are calmer with that color. The second reason is that white is much cooler in summer than other colors, and since the suit is usually made out of thick and durable material, you want any help you can get with keeping cool.

Back in Chapter 1, when we outlined bee suits, we stated that the option is to get either a full suit or just a jacket. The advantage of a suit is that it will have gatherings at the arms and legs that keep you protected from bees getting at your skin, but if you get separate pieces, just make sure that any loose ends are tucked into each other to make it extra secure. The jacket sleeves should be put into the gloves, and your pants should be covered by the tops of your boots. Tall boots, like rain boots, are best since as you move frames around, bees can fall to the ground and then crawl up your legs—with the boots on, they cannot get to your skin.

Smoker

A smoker is an absolutely valuable tool, but it does need some knowledge on how to use it properly. As we have stated before, smoke makes your bees more docile because it makes them believe that a fire is nearby, so they gorge on honey, which makes them slow and less likely to attack.

But how does the smoker itself work? The main body of the smoker is taken up by a fire chamber or firebox, where you burn fuel that produces the smoke that escapes through the nozzle. A bellows is attached to the fire chamber and allows you to pump in fresh air to get the fire burning stronger.

It is best to use three parts of fuel to get the fire burning: some tinder that has a short burning time, kindling that burns slightly longer than the tinder, and the main fuel, which will burn the longest. There are numerous options for these, ranging from natural and everyday products (newspaper, pine needles, and wood chips) to specially bought products.

Whatever fuel you decide on, make sure that you are lighting it the proper way. You want your fire to burn for more than ten minutes without needing to use the bellows when you are using it in your hives, so it is necessary to get a strong burn going before opening the hive. The steps you need to take are to:

1. Make sure you have all three parts of the fuel at hand.

2. Light the tinder and then put it into the fire chamber, placing it into the bottom with your hive tool. Make sure you leave lots of open space around the tinder—give the bellows a couple of pumps to get some extra air on the fire.

3. Add the kindling to the fire, carefully pumping the bellows. Once the first amount of kindling has lit, carefully add some more, making sure that you are giving the fire plenty of room to breathe.

4. Keep adding kindling until you no longer need to pump the bellows—at this point, push the fuel down more compactly, filling up half of the fire chamber, then add your main fuel.

5. Continue using the bellows until a steady fire is burning and smoke is constantly coming out. Once the fire is burning well, leave it for a minute to make sure that it will stay burning, and then you are good to use the smoker in the hives. Make sure that the smoke is cool to the touch.

Once you are ready for the inspection, make sure that you first smoke the entrance of the hive and then leave it for a few moments before lifting the lid and smoking the inside. As you move through the different boxes, you can give a quick puff of smoke, but make sure that you do not use too much, as that can leave soot buildup on the comb. If sparks start coming out of the top of the smoker, it means that the fuel is running low, and you need to add in some more. Once done, let the fire burn through all the fuel, let the embers cool, and then dispose of them.

Smoker tip! If you get stung, direct some smoke onto the area, as that helps mask the pheromones the angry bee left behind, making it less likely that others will attack.

Outside Threats

Now that you know some beekeeper mistakes to avoid, it is time to look at common external threats to your hive. These problems are harder to plan for, but knowing what to do in case they strike will save you time and stress if they do happen.

Excess Moisture

If there is too much moisture in your hive, condensation can form, which is not good for your bees. They do need some water for proper maintenance, but too much forms cold drops that cling to all the surfaces in the hive. This excess water then cools the hive down more than what your bees want, and the bees themselves get wet, which makes it harder for them to survive.

The seasons where you need to watch out for condensation are winter and spring. Dealing with excess moisture in winter is a little easier since you can just open the top lid or your ventilation holes for a short period to help get some more airflow through the hive.

Spring is a little harder since the water buildup is likely a result of the wet weather and water clinging to foraged pollen. If you are seeing lots of water buildup during the rainy months, consider installing moisture control items, such as a moisture board or an empty hive box filled with absorbent material.

Pest Control

Probably one of your main concerns is how to deal with all the different kinds of pests that can plague your hive. One common way to help deal with all pests is to put your hive on a stand so that it is off the ground. That helps to make it harder for them to crawl into the hive. Another tip is to properly discard the comb as you remove it from the hive and not let it fall to the ground. Try to keep as much honey off the ground as well, since that sugary sweetness will attract pests, who will then try to get into the hive for more.

Some chemical methods can be used to deal with pests, but make sure that you do your research first. Many insect killers will work against the invaders, but will also kill your bees, so use sparingly and only the kinds that are least harmful to your bees (and yourself). Always wear proper protection when spraying chemicals.

Ants

Depending on the area you live in, ants can either be a major or minor concern. Luckily, if your colony is well established and strong, they will have an easier time driving the ants away. It is with new hives or ones that have gone through some hardship that you need to be most vigilant. Not only will they steal honey, but also larvae, which will further weaken the hive.

There are a few steps to take to prevent ants from becoming an issue. First, know what ants are in your area and during what season they are most active. This will help you know when you need to be observing more closely. Second, keep the areas around your hives free from tall weeds, as that gives fewer places for the ants to hide, so you can see them easily as they try to get into the hive.

Observing how they get into the hive is important because ants, like bees, are pheromone-driven. This means that they leave a lovely trail to follow, so finding that trail and disrupting it can stop their procession into the hive.

The unfortunate thing is that similarities between bees and ants do not end there—if you try chemical methods of killing the ants, it is just as likely to kill your bees. So, if the ants are already in your hive, the best option is to physically scrape them out. Of course, this is hard to do since they are so small.

It is better to prevent them from getting inside at all. A common method that beekeepers use is to create moats around the legs of hive stands. Ants cannot swim, so putting some water (mixed with a little oil) around the stand means that they cannot climb them. This can be as simple as putting cans of water (made slick by some means) under the hive stand legs, covered by disposable cups so that the bees do not fly into the moats.

Other options would be to slick up the hive legs with motor oil or grease so that the ants cannot climb them, though this can get messy and needs to be reapplied after a few months. Another option to put directly onto the legs is Tanglefoot, a natural glue product often used on fruit trees that traps the ants in place. Like the motor oil, this can be messy to work with and needs to be reapplied. A more natural approach is to create barriers of ground cinnamon around the hive legs, which deters the ants but needs to be reapplied more frequently since it is easily blown away.

Beetles

Hive beetles (small and large beetles) are pests that can do a lot of damage to your hive and bees. Detecting and getting rid of them early is crucial.

Small hive beetles are dark brown or black and are only five millimeters long! Despite their small size, they can cause a lot of destruction—they burrow through the comb, defecating as they go, which ferments and discolors the honey. They also have a taste for bee larvae.

Once in a hive, they are everywhere! The bees will try to fight back and drive the small beetles into crevices that they then seal off with wax.

Unfortunately, the beetles then lay their eggs in these sealed areas, which then hatch and start eating through the comb.

Large hive beetles are also black but much larger, around 20 to 23 millimeters. Unlike small beetles, these do not lay their eggs in the hive but in the ground under and around it. However, they cause damage by getting it as adults and eating through your brood. To keep large beetles out of your hive, make sure that your entrance is small enough that they cannot get through.

Treatment for beetles includes the use of different kinds of bee-safe chemicals to be put on the ground or around the hive legs. Other options are to shine a light near the larvae-infested area, which gathers them together so you can collect them all. You can also take infested frames out of the hive and hang them near chickens if you have them. The beetle larvae will drop out of the frames and get gobbled up by the chickens. Another option is to freeze the frames for four days, which will kill the larvae.

Mites

The most damaging pests a beekeeper has to deal with are these mites, as they threaten the survival of a hive once they become established. They attach themselves to adult bees and suck their blood and lay their eggs in brood cells, where their larvae feed off bee babies, infecting them with viruses and weakening and even killing them.

Mites are difficult to detect and require regular testing to find them—waiting too long lets the mites get stronger as they slowly kill your hive. There are several methods of detecting and dealing with mites, but all take time and effort and have pros and cons. When picking one, make sure that it is a method that you can use as quickly and least invasively as possible.

Sugar Dusting

Have some powdered sugar in a container, and during an inspection, take a group of adult bees from at least three frames. Quickly place those bees into the container, sealing it with air holes, so they cannot escape. Then, gently shake or roll a couple of times—this coats the bees in the sugar, which does not hurt them.

Have a mesh screen or strainer on hand. After a few minutes and several coatings, since the more sugar on the bees, the more effective it is, sift the sugar through the mesh into another container filled halfway with water. Then leave the bees near the hive entrance so they can fly home.

In the water container, the sugar will dissolve, but it may need a stir. The mites will then be revealed. You may need to look very closely, as the mites are very small.

Drone Uncapping & Mite Trapping

Mites prefer to suck the blood of drones—no one knows why exactly, but it makes knowing where they might be a little easier. By examining the drone brood or having a specialty drone-trapping frame, you can find and get rid of some of the mites.

During an inspection, remove the frames that house the drone brood. This is when the special drone frame comes in handy. You need to make sure to wait for the brood to be capped for this to work. With the capped brood frame out of the hive, scrape away sections of the cap, examining the exposed pupae for the reddish-brown mites.

Once you have determined that they are present, either freeze the frame or scrape it clean. Both of these do kill the brood as well as the mites, but it is a better option than raising a batch of mites along with your drones. Make sure to clean it thoroughly before placing it back into the hive.

Having dedicated mite traps comes in handy here—with one or two frames, the majority of the mite population will be in these broods. Monthly cleaning of these frames then removes most of the mites before they start affecting the rest of the hive.

Medical Intervention

If all else fails, there are medical or chemical methods (miticides) that you can use against the mites. This should be your last choice since there has been growing evidence that using medicines as a preventative measure makes the mites more resistant (Blakiston, n.d.). It is also important to note that if you use medications when there is honey present, that honey cannot be consumed by humans, though it is safe to feed that honey back to your bees.

Apistan® strips are laced with the miticide fluvalinate. Place multiple strips throughout the hive, and the bees will brush against them, transferring the chemical throughout the hive. It will not harm the bees but will kill the mites.

Formic acid gel packs can also be used, though they are caustic and can be hard to handle. Instead, try Apiguard—a natural product that contains *thymol* (a substance from thyme). Apiguard is also in gel packets that you place on top of the hive frames, though these are time-released to ensure that the correct amount of miticide ends up in the hive.

Large Pests

Along with the smaller pests, you also need to think about the possibility of larger pests and predators giving you trouble. The threat of most of these creatures depends largely on where you live, but even in a city center, they have a habit of creating problems.

Of these largest pests, bears are the most worrisome. These are not Winnie the Pooh looking for honey pots—they are large wild animals that will tear apart your hive to get at the honey and larvae inside. If you are visited by bears, you will most likely need to fully replace your hives and bees and also contact local conservation officers. Once they know where a hive is, the bear will likely come back again for another snack. Unfortunately, the only guaranteed way to keep bears away from your bees is to install an electric fence as protection. This is not the cheapest or easiest option, but much better than having bears visit and wreck things.

In an urban setting, it is more likely that you need to worry about skunks and raccoons. Skunks love to snack on insects, so when they target a hive, they will be going after the bees and not the honey. Active at night, they will scratch at the hive until the bees come out so they can eat them. Having your hive on a stand prevents the smelly buggers from eating your bees since they have to stand up to get at the hive, which then exposes their soft underbellies. The guard bees then target that area, making the skunks think twice about whether the snack is worth the pain. Other options for keeping them away are:

- Poultry fencing (since skunks do not climb)
- Live traps
- Hammering some nails into a board and laying it on the ground as a barrier

Just make sure that you do not step on it yourself!

Raccoons will climb over any fencing in place to get on top of your hive. Once there, they can open the lid and scoop out the honey and bees, so the best bet to keep these bandits away is placing heavy rocks on top of the hive. Simple and inexpensive!

Another concern is mice—these horrid little things will not hurt your bees, but if they get into the hive, they will damage the comb. In the winter, they will make nests inside the hive, and in doing so, they will chew through the comb and possibly the frames as well. To avoid mice, once it starts cooling down, install a metal grate entrance reducer or mouse guard that has spaces large enough for your bees to get through but too small for mice.

The last 'pest' that you need to worry about is—children. Kids are very curious, and if you do not take the proper precautions to keep them away from your hives, it is likely to result in stings or even hive damage. It is best to keep your bees and children safe by child-proofing the apiary area as much as possible. To do this:

- Teach children how to act around bees—i.e., try not to get scared or swat them.

  o Also, teach them to stay calm and get the help of an adult in removing the stinger if they get stung.

  o Have rules for when they can and can't go to the hive—if their activity near the hives is monitored, there is less chance of any accidental destruction taking place.

- Warn neighbor families that bees will be around—you do not want to take the chance that a local child who is allergic to bees will get stung.

- Keep children away from the hive in late summer when the bees are most active and protective of their honey.

- If they are helping you with the hive, make sure that they are wearing the proper protection just the same as you are.

  o Only give them age-appropriate tasks.

- Have a good amount of distance between your hives and the areas where children play.

All of these will help to protect the children and the bees as well. They can co-exist—it just needs to be done carefully.

* * *

While not an extensive list of all the potential roadblocks and hardships you may face as a beekeeper, at least now you have an idea of some of the most common ones. Being forewarned is forearmed, after all.

But to stay truly ahead of any pitfall, you need to continue growing your knowledge. You need to always keep learning and expanding your knowledge base—if you stagnate, so will your hive, and eventually, it will all taper off into nothingness.

You can Google this term to learn more:

- Getting Rid of Hive Beetles

Chapter 6:

A Growing Operation

All of the information presented in this book has come down to this point. Following what has been written, you should be progressing from a beginner to an expert beekeeper. This chapter is all about expanding your operation and possibly turning it into a profit as well!

It does not matter when, or if, you move on to this stage—you can do this after the first year, after your first hive is fully established, or whenever. The only important time constraint is making sure that you are comfortable dealing with the changes.

From here, we will cover taking it all to the next level—how to be prepared for adding additional hives and how to deal with the expansion that needs to happen within the hives. As a bonus, the final part of this chapter outlines how to begin making beekeeping a profitable endeavor!

Hive Growth

As your colony gets more established, its numbers grow. This means that the space they need also grows, and as a dedicated beekeeper, you need to provide that space unless you want your bees to swarm away.

Starting from the very beginning of installing your bees, you probably have only a few boxes in your hive. This is good since the smaller colony will be able to stay warm in a smaller space better than a large one. This is good to start with, but as they begin to be more productive, you need to add specific spaces for the queen and brood and honey production. This results in having separate boxes, such as the honey super box, or maybe even multiple honey super boxes! After all, you do not need to limit yourself to just one of any kind.

The number of boxes you need in your hive depends on the size of your colony. As the number of workers increases, that means that the working space increases as well. Keep a close eye on your frames and see how many are currently being used by the bees.

No matter the number of frames you have within the hive, when there are only three or four empty frames left, it is time to consider adding another box with additional frames. This will give the bees room to keep building comb on the older frame as well as expand upwards.

Keep adding extra boxes as needed, stopping when it reaches the point where inspections and management are no longer easy or possible. At that point, you are going to have to introduce the bees into a new hive altogether or risk them either dying or swarming to locations unknown.

Honey Extraction

No matter the end goal you want with your honey, whether that's a delicious personal snack or a jar to sell, you still have to harvest it from the hives. Before going into the steps needed, make sure that you know how much honey you need to leave for your bees and that you have a honey extractor that you can use. You can work without the extractor, but it is much messier and harder to do—if you do not have a personal one, see if anyone in your club has one you can borrow.

Also, to prepare for honey harvesting, you can put a bee escape in the access hole between the brood and honey boxes. These pieces make it so the bees can only travel in one direction—placing the escape 24 hours before harvest will make it so that there will be fewer bees in the honey super that you will need to deal with. If you do not have that much time, you can use a product like Bee-Quick, which releases a smell that bees do not like, driving them away.

Once everything is planned and prepped, you start the process the same way you would an inspection—with your suit on and the hive smoked. Once that is done, and you have gotten into the honey super, remove the frames that you want to harvest, at least three-quarters capped. These frames then get fully removed from the hive and taken away. Store them in at least a room temperature area to begin softening them up.

Choose your harvesting location carefully—no matter what, it will get at least a little messy, so it is best to have that all sorted and set up before working with the honey. Honey extraction (when using an extractor) then happens in these steps:

1. Remove wax caps off the comb. This is easily done with a heated electric knife. Starting from the top, carefully cut away the wax, leaving as much honey and comb behind as possible.

   a. These wax caps are your stored beeswax, which can be used and sold as well.

2. Place the cleaned frames into the extractor, trying to balance out the weight of frames on either side. Start the extractor spinning with a bucket to collect the honey underneath the spout.

   a. Most extractors only remove honey from one side at a time, so you will need to flip the frames to make sure both sides are emptied.

   b. How fast the honey flows depends on the machine and the temperature of the honey—the warmer it is, the faster it moves.

3. Once all the honey is off the frames, return them to the hive. The bees will then clean them the rest of the way.

4. For the honey, leave it for a while in the buckets or jars so that all air rises to the top and escapes. After that, it is good to be bottled and stored.

And that is it! It is a pretty simple process and only requires a few different steps from a normal inspection. If you can't get access to a honey extractor, you can also hang the frames in a warm space (not direct sun) and let the honey drip into some buckets, but that will take longer.

There are two other methods of honey harvest—comb honey and the crush-and-strain method. Comb honey is perhaps the simplest since instead of cutting the caps off, you just cut the comb up altogether and package it with the honey. This may not sound good, but apparently, it is very tasty and the desired bonus by some people. The crush-and-strain method is very similar to comb honey, though after cutting the comb into pieces, it is squeezed or strained to remove as much of the honey from the wax as possible. The leftover wax then gets discarded.

BONUS—Turning Beekeeping Into a Business

Whether you have arrived at this stage because it was your intended purpose or because you have excess honey and wax, you can make some extra cash being a beekeeper. This final part of the book outlines what you need to consider when wanting to make a profit. It is time to take your hobby to the next level!

The first thing you need to consider when turning your hobby into a business is whether or not doing so is legal. To find this out, return to your research about your local zoning laws and homeowners association rules. You may already have this information from your research into the laws about keeping bees in the first place, so it could be as simple as a refresher to find this out. If you do not already have the information, use the contact you made before to help find out these specific guidelines. When doing so, make sure also to check out local laws about selling honey and wax to make sure that you have them up to code.

After that, properly plan out your business model before you start selling—this includes figuring out operational costs, deciding where and what you are going to sell, and to whom you are going to sell. Operational costs should not greatly exceed what you already have invested, but make sure that you include any business or operating licenses you will need. Look into whether you need any extra insurance since that could be a bigger financial strain than you anticipate.

During the planning stage, it is important to look back over the records you have kept. The most important of these records are the expense reports since those will be the basis for calculating what price to charge for your services. If you are planning on turning your beekeeping into a business, write down all your expenses right from the start. Hire a professional accountant unless you happen to be one yourself, and ask them to help with the finances, as they will know what deductions you can claim come tax season.

An accountant will also help you maneuver what taxes and fees you will need to pay as a business. They can help you get the proper financial paperwork in place at all stages of your honey business, and that will help you avoid tax-related pitfalls. After all, no one wants unexpectedly to owe more taxes than they thought.

To be best prepared, talk to the members of your club and professional beekeeping associations. They are likely the most up-to-date source of information you have that will help you keep all your bases covered. Many of them are probably also selling their products, so they will know the laws and administrative duties you will need to do and the best places to sell and advertise your products.

They will probably tell you not to jump into business right away. It is best to give yourself time to get used to the duties of a beekeeper before adding the extra stress of trying to turn a profit.

Once you feel confident in your abilities as an apiarist and have made your business plan, make sure that you have planned everything else out as well. Depending on the amount of honey you want to sell, you likely want to buy a honey extractor if you do not already have one. This can be a several hundred dollar cost, or a several thousand dollar one, depending on where you buy it. But the value is well worth it since you will be able to get more honey from the combs much faster than any other method.

Thinking about the honey, you also need to decide how it is going to be packaged. Regular canning jars and lids can work, but if you want to avoid doing the lettering and the look of a typical jar, find a local supplier who has more options. Avoiding the raised lettering is also really good for your labels since they stick better on a smooth surface. Make sure that you check local laws for these labels; most states and provinces have specific laws about what can and can't be on a honey label. These are food items, so health and safety regulations apply and can't be something you skip over.

The last things you need to consider are the more fun specifics of your business. What are you going to sell? Honey is the first obvious choice, but you can also sell beeswax straight or mixed in other products, as well as your bees' service as pollinators. This could be done by having someone, say a farmer, rent the number of your hives they want. They would then be paying you rental fees and the cost of labor and care that you would need to provide in the form of regular inspections. If you live further away, you may even consider charging for travel time.

If you have enough bees and to make your business bigger, you can also consider adding the bees themselves to your for-sale list. You can make your own bee nuc colonies by separating some frames from the rest of the hive. Let them make a new queen and leave them to grow strong over the winter. In the spring, the small colony is ready to be sold to someone new.

There are so many different options for what you can sell. To give yourself the strongest business, diversify your products. Reliance on just one or two items is likely to leave you with less profit than you would like.

After deciding what you will be selling, figure out who you will be selling to and how you will accomplish that. Are you going to sell at local farmers' markets? Rely on word of mouth? Become a vendor at a local artisan store? Have an online store? There are endless possibilities, so make sure to weigh the pros and cons of them all. The great news is that the demand for fresh, locally made products has never been higher, so turning your hobby of beekeeping into some extra cash should be a bee-reeze!

Conclusion

You did it! You finished the book and are well on your way to becoming a successful beekeeper in the next year. Congratulations! Getting the proper knowledge and information is crucial, and it should now be ingrained into your mind.

You may have picked up this book for so many different reasons—you wanted to get a fun but not physically demanding outdoor activity, you are trying to return to the land and away from corporate life, or are wanting to start a new business. No matter what your reasoning, we hope that when you put this book down, you will feel much more confident about achieving your goal.

To be successful in the end, make sure that reading this and other books is not your final step on the knowledge journey. Creating a firm foundation of knowledge and support is crucial to being a beekeeper, and that is why joining a beekeeping club will significantly benefit you. From there, you can draw on the support and knowledge of more experienced apiarists, take courses to help solidify your knowledge, and in time, pass that knowledge on to someone new. Beekeepers of all levels need support and people who share in their passions, and a club is an excellent way to form a community.

As a move to get back to the land, you may be reading this from your home in the city. There is a steady rise in urban beekeepers, so you are not alone if that is something you are planning to do. Make sure that you pick the perfect spot for your hives, be aware of the flight path, have access to food and water, and allow ease of your access to the hives. Optimizing location before buying your bees is the best for ultimate success.

Once your location is chosen, make sure that you know everything you need to set up your hive. This includes choosing the style and material of your hive, giving yourself enough time to get things organized, and how and where to buy your bees. There are numerous options during this stage, so make sure you take your time and have fun with it all.

Getting your bees into your hive is the next important step. This means a proper placing of the queen and knowing how much to feed the bees and when to do so. Once they are housed, make sure that you do regular inspections so that you keep on top of all proper maintenance and care that your bees may need.

With that in mind, there are potential pitfalls that can befall even the most experienced beekeeper. Being aware of these problems and keeping an eye out for them so you can stop them early is much easier than intervening once the issue has taken root. These include making sure that your hive continually has a queen, not harvesting too much honey, using beekeeping equipment properly, and preventing pests from making a mess of your hive. Being aware of these problems can help you stay ahead of the curve and avoid negative consequences.

It is a lot of work and time, but it is worth it in the end when you have that first jar of delicious honey! When you hold up that amber liquid, you should be incredibly proud of yourself. You can then choose to try and sell that honey as well as other products from your bees or keep it all for yourself. Both have merits, and, as we have said, it is a personal choice and entirely up to you. Maybe consider making some honey butter to spread over freshly baked bread? We promise you, the taste is almost addictive, and you will want to do it again and again.

You have the tools—get on out there and do great!

References

Anthony. (n.d.-a). *Essential equipment for beekeeping: 6 things you need to get started.* Beekeeping 101. Retrieved May 31, 2021, from https://www.beekeeping-101.com/essential-equipment-beekeeping-6-things-get-started/

Anthony. (n.d.-b). How to use a bee smoker. Beekeeping Insider. https://beekeepinginsider.com/how-to-use-a-bee-smoker/

Apiguard. (n.d.). Apiguard—Frequently asked questions. Retrieved June 15, 2021, from https://www.vita-europe.com/beehealth/wp-content/uploads/Apiguard.pdf

Arcuri, L. (2020, June 30). Inspecting a honey bee hive. The Spruce. https://www.thespruce.com/inspect-your-honey-bee-hive-3016536

BBC Gardener's World Magazine. (2019, July 8). How to make a bumblebee pot. BBC Gardeners' World Magazine. https://www.gardenersworld.com/how-to/grow-plants/how-to-make-a-bumblebee-pot/

BBC Gardeners' World Magazine. (2019, August 5). How to make a bee-friendly garden. BBC Gardeners' World Magazine. https://www.gardenersworld.com/plants/how-to-make-a-bee-friendly-garden/

Bees & Your HOA: *What Rights do homeowners have to keep bees in their backyards?* (2020, August 1). LeaseHoney. https://leasehoney.com/2020/08/01/bees-your-hoa-what-rights-do-homeowners-have-to-keep-bees-in-their-backyards/

Blackiston, H. (n.d.-a). *11 tips for extending the life of your beehive equipment.* Dummies. https://www.dummies.com/home-garden/hobby-farming/beekeeping/11-tips-for-extending-the-life-of-your-beehive-equipment/

Blackiston, H. (n.d.-b). *Pick the perfect location for your beehive.* Dummies. https://www.dummies.com/home-garden/hobby-farming/beekeeping/pick-the-perfect-location-for-your-beehive/

Blakiston, H. (n.d.-a). *How to control a varroa mite problem in your beehive.* Dummies. https://www.dummies.com/home-garden/hobby-farming/beekeeping/how-to-control-a-varroa-mite-problem-in-your-beehive/

Blakiston, H. (n.d.-b). *How to keep larger animals out of your beehive.* Dummies. https://www.dummies.com/home-garden/hobby-farming/beekeeping/how-to-keep-larger-animals-out-of-your-beehive/

Blakiston, H. (2016). *How to choose lumber for your Beehive.* Dummies. https://www.dummies.com/home-garden/hobby-farming/beekeeping/how-to-choose-lumber-for-your-beehive/

Built, B. (n.d.-a). Buying your first hive. Bee Built. https://beebuilt.com/pages/buying-your-first-hive

Built, B. (n.d.-b). Top bar hives. Bee Built. https://beebuilt.com/pages/top-bar-hives

Built, B. (n.d.-c). Warre hives. Bee Built. https://beebuilt.com/pages/warre-hives

Burlew, A. : R. (2020, September 2). *Sun and shade for bees: What is the right mix?* Backyard Beekeeping. https://backyardbeekeeping.iamcountryside.com/hives-equipment/sun-and-shade-for-bees/

Burns, D. (2011, January 17). *How many hives should I start with?* Mother Earth News. https://www.motherearthnews.com/homesteading-and-livestock/how-many-hives-should-i-start-with

Burns, S. (2017, March 19). *How to set up your first beehive.* Runamuk Acres Conservation Farm. https://runamukacres.com/how-to-set-up-your-first-beehive/

Casey, C. (2018, February 12). *Bees need water: Establish water sources in late winter to keep them out of the pool in summer.* ANR Blogs. https://ucanr.edu/blogs/blogcore/postdetail.cfm?postnum=26345

Caughey, M. (2016, January 21). *9 tips when selecting honeybees.* Backyard Bees. https://www.keepingbackyardbees.com/9-tips-when-selecting-honeybees/

Charlotte. (2020, December 31). *How to provide drinking water for bees.* Carolina Honeybees. https://carolinahoneybees.com/beauty-bee-farm/

Charlotte. (2021a, March 17). *Choosing the best types of honey bees.* Carolina Honeybees. https://carolinahoneybees.com/types-of-honey-bees/

Charlotte. (2021b, April 18). *Beekeeping business - starting a bee farm.* Carolina Honeybees. https://carolinahoneybees.com/start-a-beekeeping-business-from-scratch/

Charlotte. (2021c, June 8). *How to clean beeswax the easy way.* Carolina Honeybees. https://carolinahoneybees.com/processing-beeswax-cappings/

Choosing a beehive: Buy or build? (2016, March 19). Paris Farmers Union. http://blog.parisfarmersunion.com/2016/04/choosing-beehive-buy-or-build.html

Cohenour, C. (2008, January 5). Repairing hive bodies. WV Beekeeper - Cass Cohenour. http://wvbeekeeper.blogspot.com/2008/01/repairing-hive-bodies.html

Contributor. (2021, April 11). Bee bucks – The cost of beekeeping. Backyard Beekeeping. https://backyardbeekeeping.iamcountryside.com/hives-equipment/bee-bucks-the-cost-of-beekeeping/

Creating a beekeeping business plan - guide and template. (2010). BuzzAboutBees.net. https://www.buzzaboutbees.net/beekeeping-business-plan.html

David. (2017, November 24). Principles and practice. The Apiarist. https://www.theapiarist.org/principles-and-practice/

Drone uncapping background. (n.d.). Retrieved June 15, 2021, from https://beeaware.org.au/wp-content/uploads/2014/03/Drone-uncapping.pdf

Edmondson, R. (n.d.). *How to harvest honey from a beehive.* Dengarden. https://dengarden.com/gardening/How-to-Extract-Honey-from-a-Beehive

Edmondson, R. (2021, June 3). *How to install a package of bees in a new hive.* Dengarden. https://dengarden.com/gardening/How-to-Install-A-Package-of-Bees-in-a-New-Hive

Feeding honey bees to prevent starvation. (2021, January 21). Agriculture.vic.gov.au. https://agriculture.vic.gov.au/livestock-and-animals/honey-bees/health-and-welfare/feeding-honey-bees-to-prevent-starvation

Forest Farming. (2015, October 23). Hive Placement. Www.youtube.com. https://youtu.be/SEMYPO6ozSk

Foust, S. (2018, August 27). DIY bee fountain. Thebeebox. https://www.thebeebx.com/single-post/2018/08/27/diy-bee-fountain

Fun fact - Why do beekeepers wear white? (2017, September 2). Manuka Vet. https://www.manukavet.com/blog/post/20394/Fun-fact-Why-do-beekeepers-wear-white/

Gardener, A. (2015, May 6). Bee predators in beekeeping. Blain's Farm & Fleet Blog. https://www.farmandfleet.com/blog/bee-predators-beekeeping/

Government of New Brunswick, C. (1996, July 10). *Managing honeybee hives for the pollination of wild blueberries.* Www2.Gnb.ca. https://www2.gnb.ca/content/gnb/en/departments/10/agriculture/content/bees/managing.html#:~:text=The%20number%20of%20hives%20necessary

Hadley, D. (2019, October 10). Why do bees swarm? ThoughtCo. https://www.thoughtco.com/why-do-bees-swarm-1968430#:~:text=Bees%20Swarm%20When%20the%20Colony%20Gets%20too%20Large&text=Just%20as%20individual%20bees%20reproduce

Helpful beekeeping website links. (n.d.). Piedmont Beekeepers Association. Retrieved May 31, 2021, from https://www.piedmontbeekeepers.com/helpful-links

Holly. (2019, August 6). *Is beekeeping legal in my city?* Complete Beehives. http://completebeehives.com/is-beekeeping-legal-in-my-city/

Holly. (2020a, March 17). *How far apart should beehives be placed?* Complete Beehives. http://completebeehives.com/how-far-apart-should-beehives-be-placed/

Holly. (2020b, August 15). *What is the best beehive stand height?* Complete Beehives. http://completebeehives.com/what-is-the-best-beehive-stand-height/

Home and Garden, P. T. (2021). *5 ways bees are important to the environment.* Premier Tech Home and Garden. http://www.pthomeandgarden.com/5-ways-bees-are-important-to-the-environment/#:~:text=As%20pollinators%2C%20bees%20play%20a

How much time does beekeeping require. (2019, July 3). The Bee Store. https://thebeestore.com.au/blogs/bee-blog/how-much-time-does-beekeeping-require

How to create a pollinator-friendly garden. (2017). David Suzuki Foundation. https://davidsuzuki.org/queen-of-green/create-pollinator-friendly-garden-birds-bees-butterflies/

How to get rid of hive beetles – The beginner's guide. (2019, May 22). BeeKeepClub. https://beekeepclub.com/how-to-get-rid-of-hive-beetles-the-beginners-guide/

Kearney, H. (2016, May 31). *How to protect your bees from ants.* Beekeeping like a Girl. https://beekeepinglikeagirl.com/how-to-protect-your-bees-from-ants/

Kearney, H. (2020, August 17). *How to tell if your hive is queenless.* Flow Hive US. https://www.honeyflow.com/blogs/beekeeping-basics/queenless-hive

Lesa. (2014, April 30). *8 honey bee hive inspection tips.* Better Hens & Gardens. https://www.betterhensandgardens.com/honey-bee-hive-inspection-tips/

McElroy, S. C. (2016, December 14). *Ventilation: It's complicated keeping backyard bees.* Keeping Backyard Bees. https://www.keepingbackyardbees.com/ventilation-its-complicated/

Mercedes, K. (2018, December 7). *5 ways joining a bee club makes you a better beekeeper.* Hobby Farms.

https://www.hobbyfarms.com/5-ways-a-bee-club-will-make-you-a-better-beekeeper/

Mortimer, F. (n.d.). *Suburban bees: How to keep bees in residential areas.* Pollinator.cals.cornell.edu. https://pollinator.cals.cornell.edu/master-beekeeper-program/meet-our-master-beekeepers/suburban-bees-how-keep-bees-residential-areas/

Nesset, J. (2016, January 18). *Can you keep bees with kids?* Murdoch's Blog: The Dirt. https://blog.murdochs.com/can-you-keep-bees-with-kids/

Nickson, J. (2019a, July 3). *How to winterize a beehive: A beekeeper's guide to preparing bees for the winter.* Honest Beekeeper. https://honestbeekeeper.com/how-to-winterize-a-beehive/

Nickson, J. (2019b, August 11). *How much time does beekeeping take? It's more than you think.* Honest Beekeeper. https://honestbeekeeper.com/how-much-time-does-beekeeping-take-its-more-than-you-think/

PerfectBee. (n.d.). *Finding beekeeping clubs and mentors.* Www.perfectbee.com. https://www.perfectbee.com/learn-about-bees/about-beekeeping/beekeeping-clubs-and-mentors

Plant Health Australia. (n.d.). Sugar shaking BACKGROUND. Retrieved June 15, 2021, from https://beeaware.org.au/wp-content/uploads/2014/03/Sugar-shaking.pdf

Ploetz, K. (2013, May 8). *Dear modern farmer: How do I legally start an urban bee hive?* Modern Farmer. https://modernfarmer.com/2013/05/dear-modern-farmer-how-do-i-legally-start-an-urban-bee-hive/

Poindexter, J. (2016a, June 8). *38 Free DIY bee hive plans that will inspire you to become a beekeeper.* MorningChores. https://morningchores.com/beehive-plans/

Poindexter, J. (2016b, November 20). *15 essential beekeeping equipment every beekeeper can't live without.* MorningChores. https://morningchores.com/beekeeping-equipment/

Queenlessness in Your Hive - PerfectBee. (n.d.). Https://Www.perfectbee.com. https://www.perfectbee.com/a-healthy-beehive/inspecting-your-hive/queenlessness-in-your-hive

Rose, S. (2016, November 1). Build a bug hotel. Garden Therapy. https://gardentherapy.ca/build-a-bug-hotel/

Rusty. (2010a, March 2). Pollen collection by honey bees. Honey Bee Suite. https://www.honeybeesuite.com/pollen-collection/

Rusty. (2010b, May 19). *Reduce varroa mites by culling honey bee drones.* Honey Bee Suite. https://www.honeybeesuite.com/reduce-varroa-mites-by-culling-honey-bee-drones/

Rusty. (2010c, September 20). *Entrance reducers can annoy your honey bees.* Honey Bee Suite. https://www.honeybeesuite.com/entrance-reducers-can-annoy-your-honey-bees/

Rusty. (2011a, February 18). *How long should I feed a new package of bees?* Honey Bee Suite. https://www.honeybeesuite.com/how-long-should-i-feed-a-new-package-of-bees/

Rusty. (2011b, May 19). *Sun or shade: which is best for the bees?* Honey Bee Suite. https://www.honeybeesuite.com/sun-or-shade-which-is-best-for-the-bees/

Rusty. (2014, August 19). *How much honey do bees need for winter?* Honey Bee Suite. https://www.honeybeesuite.com/how-much-honey-should-i-leave-in-my-hive/

Rusty. (2017, June 5). *Sun is for foraging, but bees love shade.* Honey Bee Suite. https://www.honeybeesuite.com/sun-foraging-bees-love-shade/

Schneider, A. (2020, December 9). *The ins and outs of buying bees.* Backyard Beekeeping. https://backyardbeekeeping.iamcountryside.com/beekeeping-101/buying-bees-miller-bee-supply/

Screen bottom boards vs. solid bottom boards: Which is better? (n.d.). Www.perfectbee.com. https://www.perfectbee.com/blog/screen-bottom-boards-vs-solid-bottom-boards-better

7 fast solutions to get rid of ants in a beehive (For GOOD!). (2019, May 22). Backyard Beekeeping 101. https://backyardbeekeeping101.com/ants-in-beehive/

Spinks, S. (2018, November 22). Cleaning beeswax. Norfolkhoneyco. https://www.norfolk-honey.co.uk/post/cleaning-beeswax-part-1

"Start beekeeping" courses. (2020, January 17). The Apiarist. https://www.theapiarist.org/start-beekeeping-courses/

10 free DIY beehive stand plans & ideas you'll fall in love with. (2020, May 19). Backyard Beekeeping 101. https://backyardbeekeeping101.com/bee-hive-stand-plans/

The beekeeper suit: A comprehensive guide. (n.d.). BeeKeepClub. https://beekeepclub.com/beekeeping-equipment/the-beekeeper-suit-a-comprehensive-guide/

The beekeeper's calendar(n.d.). Dadant & Sons 1863. https://www.dadant.com/learn/the-beekeepers-calendar/

The Beekeepers Club Inc - events. (n.d.). Www.beekeepers.org.au. Retrieved May 31, 2021, from https://www.beekeepers.org.au/events

The Editors. (2021a, March 27). *Beekeeping 101: Should you raise honey bees?* Old Farmer's Almanac.

https://www.almanac.com/beekeeping-101-why-raise-honeybees

The Editors. (2021b, April 29). *Beekeeping 101: Choosing a type of beehive.* Old Farmer's Almanac. https://www.almanac.com/beekeeping-101-types-of-beehives

The growth and feasibility of urban beekeeping. (n.d.). Www.perfectbee.com. https://www.perfectbee.com/learn-about-bees/about-beekeeping/growth-of-urban-beekeeping

Tips for giving your beehives enough ventilation this summer. (2020, May 15). Beekeeping Resources. https://www.mannlakeltd.com/mann-lake-blog/tips-for-giving-your-beehives-enough-ventilation-this-summer/

Top 10 best beekeeping starter kits (2020). (2016, August 22). BeeKeepClub. https://beekeepclub.com/best-beekeeping-starter-kits/

Top 10 best hive stands for beekeeping (2021) – Why they are necessary. (2018, February 22). BeeKeepClub. https://beekeepclub.com/best-hive-stands/

Trujillo, T. (2019, June 17). *7 essential beekeeping equipment that every beginner needs.* Palm Pike. https://palmpike.com/7-essential-beekeeping-equipment-that-every-beginner-needs/

Varroa mites. (2016). Beeaware.org.au. https://beeaware.org.au/archive-pest/varroa-mites/#ad-image-0

Watson, B. (2019, June 13). *The real cost of beekeeping for the first year (Plus how to save!).* Backyard Beekeeping 101. https://backyardbeekeeping101.com/beekeeping-cost/

Wildlife Preservation Canada. (2021). Rusty-patched bumble bee. Wildlife Preservation Canada.

https://wildlifepreservation.ca/rusty-patched-bumble-bee/#:~:text=One%20of%20the%20most%20common

Williams, S. (n.d.). Managing hive capacity. Www.perfectbee.com. https://www.perfectbee.com/a-healthy-beehive/inspecting-your-hive/managing-hive-capacity

Williams, S. (2021). *The PerfectBee Academy online beekeeping course.* Www.perfectbee.com. https://www.perfectbee.com/academy-beekeeping-course#section-80-203966

Withers, J. (2013, October 21). *How to wrap a bee hive for cold winters.* Honey Bee Suite. https://www.honeybeesuite.com/how-to-wrap-a-hive/

Wyatt, L. J. (2015, June 19). *3 ways to harvest honey.* Hobby Farms. https://www.hobbyfarms.com/3-ways-to-harvest-honey-4/

Woodward, A. (2019, June 21). *Bees and insects dying at record rates are sign of 6th mass extinction.* Business Insider. https://www.businessinsider.com/insects-dying-off-sign-of-6th-mass-extinction-2019-2

Printed in Great Britain
by Amazon